The Handbook of Environmental Chemistry

Founded by Otto Hutzinger

Editors-in-Chief: Damià Barceló • Andrey G. Kostianoy

Volume 17

The Handbook of Environmental Chemistry
Recently Published and Forthcoming Volumes

Polyfluorinated Chemicals and Transformation Products
Volume Editors: T.P. Knepper and F.T. Lange
Vol. 17, 2012

Brominated Flame Retardants
Volume Editors: E. Eljarrat and D. Barceló
Vol. 16, 2011

Effect-Directed Analysis of Complex Environmental Contamination
Volume Editor: W. Brack
Vol. 15, 2011

Waste Water Treatment and Reuse in the Mediterranean Region
Volume Editors: D. Barceló and M. Petrovic
Vol. 14, 2011

The Ebro River Basin
Volume Editors: D. Barceló and M. Petrovic
Vol. 13, 2011

Polymers – Opportunities and Risks II: Sustainability, Product Design and Processing
Volume Editors: P. Eyerer, M. Weller, and C. Hübner
Vol. 12, 2010

Polymers – Opportunities and Risks I: General and Environmental Aspects
Volume Editor: P. Eyerer
Vol. 11, 2010

Chlorinated Paraffins
Volume Editor: J. de Boer
Vol. 10, 2010

Biodegradation of Azo Dyes
Volume Editor: H. Atacag Erkurt
Vol. 9, 2010

Water Scarcity in the Mediterranean: Perspectives Under Global Change
Volume Editors: S. Sabater and D. Barceló
Vol. 8, 2010

The Aral Sea Environment
Volume Editors: A.G. Kostianoy and A.N. Kosarev
Vol. 7, 2010

Alpine Waters
Volume Editor: U. Bundi
Vol. 6, 2010

Transformation Products of Synthetic Chemicals in the Environment
Volume Editor: A.B.A. Boxall
Vol. 2/P, 2009

Contaminated Sediments
Volume Editors: T.A. Kassim and D. Barceló
Vol. 5/T, 2009

Biosensors for the Environmental Monitoring of Aquatic Systems
Bioanalytical and Chemical Methods for Endocrine Disruptors
Volume Editors: D. Barceló and P.-D. Hansen
Vol. 5/J, 2009

Environmental Consequences of War and Aftermath
Volume Editors: T.A. Kassim and D. Barceló
Vol. 3/U, 2009

The Black Sea Environment
Volume Editors: A. Kostianoy and A. Kosarev
Vol. 5/Q, 2008

Emerging Contaminants from Industrial and Municipal Waste
Removal Technologies
Volume Editors: D. Barceló and M. Petrovic
Vol. 5/S/2, 2008

The Handbook of Environmental Chemistry Also Available Electronically

The Handbook of Environmental Chemistry is included in Springer's eBook package *Earth and Environmental Science*. If a library does not opt for the whole package, the book series may be bought on a subscription basis.

For all customers who have a standing order to the print version of *The Handbook of Environmental Chemistry*, we offer free access to the electronic volumes of the Series published in the current year via SpringerLink. If you do not have access, you can still view the table of contents of each volume and the abstract of each article on SpringerLink (www.springerlink.com/content/110354/).

You will find information about the

- Editorial Board
- Aims and Scope
- Instructions for Authors
- Sample Contribution

at springer.com (www.springer.com/series/698).

All figures submitted in color are published in full color in the electronic version on SpringerLink.

Aims and Scope

Since 1980, *The Handbook of Environmental Chemistry* has provided sound and solid knowledge about environmental topics from a chemical perspective. Presenting a wide spectrum of viewpoints and approaches, the series now covers topics such as local and global changes of natural environment and climate; anthropogenic impact on the environment; water, air and soil pollution; remediation and waste characterization; environmental contaminants; biogeochemistry; geoecology; chemical reactions and processes; chemical and biological transformations as well as physical transport of chemicals in the environment; or environmental modeling. A particular focus of the series lies on methodological advances in environmental analytical chemistry.

Series Preface

With remarkable vision, Prof. Otto Hutzinger initiated *The Handbook of Environmental Chemistry* in 1980 and became the founding Editor-in-Chief. At that time, environmental chemistry was an emerging field, aiming at a complete description of the Earth's environment, encompassing the physical, chemical, biological, and geological transformations of chemical substances occurring on a local as well as a global scale. Environmental chemistry was intended to provide an account of the impact of man's activities on the natural environment by describing observed changes.

While a considerable amount of knowledge has been accumulated over the last three decades, as reflected in the more than 70 volumes of *The Handbook of Environmental Chemistry*, there are still many scientific and policy challenges ahead due to the complexity and interdisciplinary nature of the field. The series will therefore continue to provide compilations of current knowledge. Contributions are written by leading experts with practical experience in their fields. *The Handbook of Environmental Chemistry* grows with the increases in our scientific understanding, and provides a valuable source not only for scientists but also for environmental managers and decision-makers. Today, the series covers a broad range of environmental topics from a chemical perspective, including methodological advances in environmental analytical chemistry.

In recent years, there has been a growing tendency to include subject matter of societal relevance in the broad view of environmental chemistry. Topics include life cycle analysis, environmental management, sustainable development, and socio-economic, legal and even political problems, among others. While these topics are of great importance for the development and acceptance of *The Handbook of Environmental Chemistry*, the publisher and Editors-in-Chief have decided to keep the handbook essentially a source of information on "hard sciences" with a particular emphasis on chemistry, but also covering biology, geology, hydrology and engineering as applied to environmental sciences.

The volumes of the series are written at an advanced level, addressing the needs of both researchers and graduate students, as well as of people outside the field of "pure" chemistry, including those in industry, business, government, research establishments, and public interest groups. It would be very satisfying to see these volumes used as a basis for graduate courses in environmental chemistry. With its high standards of scientific quality and clarity, *The Handbook of*

Environmental Chemistry provides a solid basis from which scientists can share their knowledge on the different aspects of environmental problems, presenting a wide spectrum of viewpoints and approaches.

The Handbook of Environmental Chemistry is available both in print and online via www.springerlink.com/content/110354/. Articles are published online as soon as they have been approved for publication. Authors, Volume Editors and Editors-in-Chief are rewarded by the broad acceptance of *The Handbook of Environmental Chemistry* by the scientific community, from whom suggestions for new topics to the Editors-in-Chief are always very welcome.

Damià Barceló
Andrey G. Kostianoy
Editors-in-Chief

Volume Preface

With this edition of *The Handbook of Environmental Chemistry* "Polyfluorinated Chemicals and Transformation Products" we aim to give an overview of the recent state of the art. Polyfluorinated chemicals (PFC) are widespread substances with effective and measurable effects to environment and economy. Topics, such as synthesis and application, analysis and degradation as well as environmental aspects, food and toxicity are spotlighted.

In this book the acronym PFC is understood as the abbreviation for all different classes of fluorinated chemicals with at least two CF_3-groups or a $CF_3(CF_2)_n$ – group with $n>0$ being implemented in the molecule. This definition also includes perfluorinated chemicals as well as fluorinated surfactants. Chemicals resulting from both, biotic and or abiotic degradation of PFC are handled as transformation products (TP). In the case of e.g., polymeric PFC the resulting TP could be PFC again. Unfortunately it was not possible to completely harmonize all abbreviations and acronyms within this book. If a chapter is dealing for example solely with perfluoroalkyl compounds, also the historical acknowledged abbreviation PFC is used for this particular chemical class.

PFC have become essential in numerous technical applications due to their unparalleled effectiveness and efficiency. The chemistry, properties, and uses of commercial fluorinated surfactants will introduce the theme.

Emphasis will be given upon compounds with improved application, environmental and toxicological properties, which are a challenge for the synthetic chemist. One chapter is dedicated to the important PFC perfluorooctanoate (PFOA), which is exemplary taken into account with regard to occurrence and uses in products.

Many PFC brought to market show limited biodegradability. The parent compound or active metabolites remain in the environment and can result in a wide spectrum of substances.

Modern analytic instrumentation enables the user to detect trace chemicals at very low concentrations but also to identify unknown compounds, such as transformation products. Various applications of modern mass spectrometric techniques as useful tools for structure elucidation are described and mass spectrometric approaches are able to reveal biotransformation products of PFC.

The environmental persistence of PFC, combined with toxic and bioaccumulative potential in some instances, has become a matter of concern. This led to the recent withdrawal of certain fluorosurfactant classes from the market. Potential health risks and biological effects cannot be excluded. Toxicological properties of fluorinated substances vary and, like the mechanisms for global distribution, are still in the process of being clarified.

To be able to predict the fate and behavior in the environment the knowledge on sorption and leaching behavior of PFC in soil is an important tool, which is addressed in a separate chapter.

Feasible for further legislative impacts is to achieve a wide data base. Thus, the remaining chapters discuss the monitoring in European surface, ground- and drinking waters, treatment options for PFC removal from drinking water, PFC in food as well as the human biomonitoring of PFC.

Idstein, Germany Thomas P. Knepper
Karlsruhe, Germany Frank T. Lange

Contents

Chemistry, Properties, and Uses of Commercial Fluorinated Surfactants

Robert C. Buck, Peter M. Murphy, and Martial Pabon

Abstract Fluorinated surfactants have been commercially available since the 1950s. The first available were perfluoroalkyl sulfonic acids. The unique properties e.g., surface tension lowering in aqueous systems, high chemical and thermal stability of these acids and their derivatives when used at low concentrations resulted in their widespread use in industrial processes and consumer uses. The most common commercially produced perfluorinated surfactants are the perfluoroalkyl acids.

Subsequently, additional commercial processes were developed for synthesis of a range of per- and poly-fluorinated surfactants whose unique properties make them largely irreplaceable in many applications. The widespread use and disposal and the high stability of the perfluoroalkyl acids, which do not breakdown readily either abiotically or biotically in the environment, has resulted in widespread presence of PFAAs in the environment. This caused commercial production to shift toward short chain alternatives and new fluorinated moieties such as the per- and poly-fluorinated ethers. Clearly, there remains a need for fluorinated surfactants in many industries to obtain the beneficial performance properties of these substances that cannot be achieved with other types of surfactants.

R.C. Buck (✉)
DuPont Chemicals and Fluoroproducts, 4301 Lancaster Pike, CRP 702-2211B, Wilmington 19880-0702, DE, USA
e-mail: robert.c.buck@usa.dupont.com

P.M. Murphy
Highway 141 & Henry Clay Blvd. E301/252, Wilmington, DE, USA 19803
e-mail: peter.m.murphy@usa.dupont.com

M. Pabon
DuPont de Nemours, S.A, 2 Chemin du Pavillon, 1218 Geneva, Switzerland
e-mail: Martial.Pabon@che.dupont.com

T.P. Knepper and F.T. Lange (eds.), *Polyfluorinated Chemicals and Transformation Products*, Hdb Env Chem (2012) 17: 1–24,
DOI 10.1007/978-3-642-21872-9_1,

The aim of this chapter is to provide an overview of the commercially relevant chemistry, properties, and uses of commercial fluorinated surfactants.

Keywords Chemical production • Fluorinated surfactants • Physico chemical properties

Contents

Abbreviations

ABS	Acrylonitrile butadiene styrene
AFFF	Aqueous film-forming foams
CMC	Critical micelle concentration
ECF	Electrochemical fluorination
EOR	Enhanced oil recovery
HFP	Hexafluoropropene
HFPO	Hexafluoropropene oxide
PBSF	Perfluorobutanesulfonyl fluoride
PDSF	perfluorodecane sulfonyl fluoride
PEM	Polymer electrolyte membrane
PFAAs	Perfluoroalkyl acids
PFCA	Perfluoroalkyl carboxylic acid
PFCAs	Perfluoroalkyl carboxylic acids

PFDS	Perfluorodecane sulfonate
PFOS	Perfluorooctane sulfonate
PFPA	Perfluoroalkyl phosphonic acid
PFPIA	Perfluoroalkyl phosphinic acid
PFSA	Perfluoroalkyl sulfonic acid
PHxSF	perfluorohexane sulfonyl fluoride
POSF	Perfluorooctane sulfonyl fluoride
TFE	Tetrafluoroethylene

1 Introduction

The surfactant universe includes a wide variety of substances from natural to synthetic that contain functional groups which provide specific performance properties for a plethora of valuable industrial and consumer uses. Fluorinated surfactants are a specific class of surfactants whose properties are derived from substitution of at least one hydrogen atom along the carbon backbone that makes up the hydrophobic part of the surfactant with fluorine [1–7]. The terms fluorosurfactant, fluorinated surfactant, and fluorinated tenside are synonyms that describe a broad and diverse group of surfactants. The extent and location of fluorine substitution in the surfactant affect the surfactant properties. For example, fluorinated surfactants with a terminal $-CF_3$ group differ from fluorinated surfactants with a hydrogen-containing terminus [2]. A polyfluorinated surfactant is one in which more than one, but not all hydrogen atoms are substituted with fluorine. The carbon–fluorine bond is very strong and the perfluoroalkyl functional group, $F(CF_2)_n-$, is both hydrophobic and oleophobic [8]. Perfluorinated surfactants represent the ultimate type of fluorinated surfactant, where *all* hydrogen bound to carbon is replaced with fluorine except those hydrogen atoms whose substitution would modify the nature of any functional groups present [9].

Fluorinated surfactants have been commercially available since the 1950s. The first available were perfluoroalkyl sulfonates (e.g., perfluorooctane sulfonate, $C_8F_{15}SO_3^-$, PFOS) and perfluoroalkyl carboxylic acids (e.g., perfluorooctanoic acid, $C_7F_{15}COOH$, PFOA) manufactured using the electrochemical fluorination (ECF) process [10]. The unique properties (e.g., surface tension lowering in aqueous systems, high chemical and thermal stability) of these acids and their derivatives when used at low concentrations resulted in their widespread use in industrial processes and consumer uses [11–13]. The most common commercially produced perfluorinated surfactants are the perfluoroalkyl acids (PFAAs):

Perfluoroalkyl acids

General name	Acronym	Structure
Perfluoroalkyl sulfonic acid	PFSA	$F(CF_2)_nSO_3H$
Perfluoroalkyl carboxylic acid	PFCA	$F(CF_2)_nCO_2H$
Perfluoroalkyl phosphonic acid	PFPA	$F(CF_2)_nP(=O)(OH)_2$
Perfluoroalkyl phosphinic acid	PFPIA	$F(CF_2)_nP(=O)(OH)$

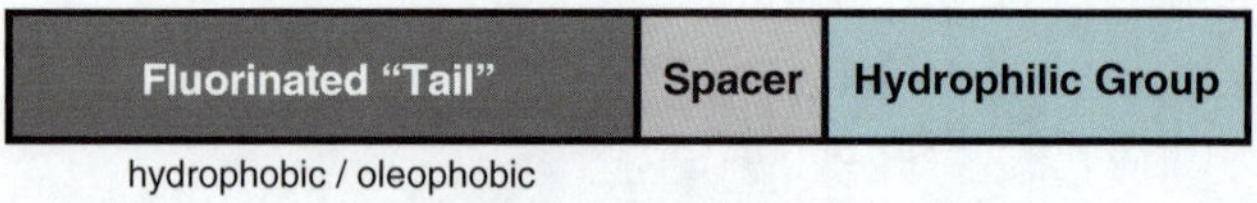

Fig. 1 Schematic of a fluorinated surfactant

Subsequently, additional commercial processes were developed for synthesis of a range of per- and polyfluorinated surfactants whose unique properties make them largely irreplaceable in many applications. The widespread use and disposal and the high stability of the PFAAs, which do not break down readily either abiotically or biotically in the environment, has resulted in widespread presence of PFAAs in the environment [14–16]. The aim of this chapter is to provide an overview of the commercially relevant chemistry, properties, and uses of commercial fluorinated surfactants.

2 Chemistry of Fluorinated Surfactants

An understanding of the chemistry of fluorinated surfactants must consider three distinct structural aspects: (1) the hydrophobic/oleophobic "tail" that contains a high proportion of fluorine, (2) the hydrophilic group, and (3) the "spacer" organic group linking these two portions of the surfactant together (Fig. 1). As with hydrocarbon surfactants, the valuable and important fluorinated surfactants include a diverse range of hydrophilic groups: (a) anionic, for example, sulfonates, sulfates, carboxylates, and phosphates, (b) cationic, for example, quaternary ammonium, (c) nonionic, for example, polyethylene glycols, acrylamide oligomers, and sugars, and (d) amphoteric, for example, betaines and sulfobetaines [2].

The practical and commercially valuable range of the hydrophobic/oleophobic "tail" of the fluorinated surfactant is limited [3, 5, 6]. Either perfluoroalkyl, $F(CF_2)_n$– or R_F-, or perfluoropolyether, $(R_FO)_n(R_FO)_m$-, groups are the hydrophobic/oleophobic portion of most commercially available fluorinated surfactants. Perfluoroalkyl-containing fluorinated surfactants generally originate from either (1) ECF with HF [4] or (2) telomerization of tetrafluoroethylene (TFE) [17]. Perfluoropolyether-based fluorinated surfactants typically originate from either (1) oligomerization of hexafluoropropene oxide (HFPO), (2) photooxidation of TFE or hexafluoropropene (HFP) [18], or (3) oligomerization of fluorinated oxetanes [19].

2.1 *Electrochemical Fluorination*

The ECF of organic compounds using anhydrous HF was the first significant commercial process for manufacturing ECF-based fluorinated surfactants [4, 10, 20, 21].

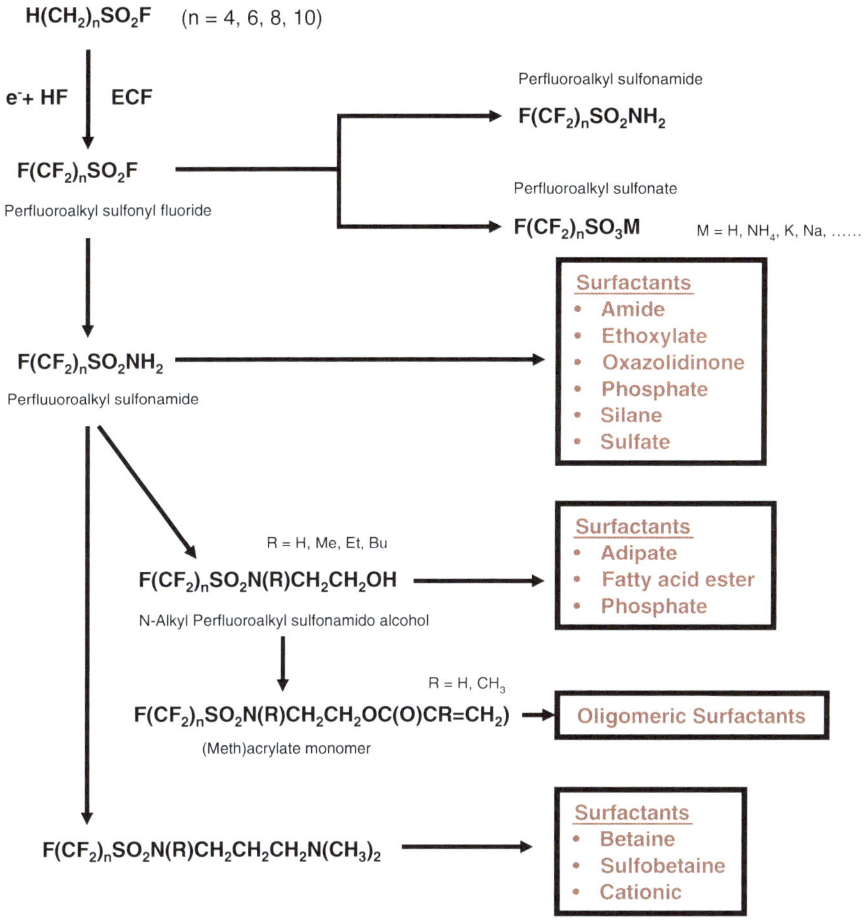

Fig. 2 Synthesis of ECF-based fluorinated surfactants

Typically, a hydrocarbon sulfonyl fluoride (R-SO_2F, for example, $C_4H_9SO_2F$ or $C_8H_{17}SO_2F$) is transformed into the corresponding perfluoroalkyl sulfonyl fluoride (R_f-SO_2F, for example, $C_4F_9SO_2F$ or $C_8F_{17}SO_2F$). The perfluoroalkyl sulfonyl fluoride is the fundamental raw material which is further processed to yield fluorinated surfactants (Fig. 2). Commercially relevant perfluoroalkyl-sulfonyl fluorides are derived from 4, 6, 8, and 10 carbon starting materials yielding perfluorobutanesulfonyl fluoride (PBSF), perfluorohexane sulfonyl fluoride (PHxSF), perfluorooctane sulfonyl fluoride (POSF), and perfluorodecane sulfonyl fluoride (PDSF), respectively. In the ECF process, fragmentation and rearrangement of the carbon skeleton occurs and significant amounts of cleaved, branched, and cyclic structures are formed resulting in a complex mixture of fluorinated materials of varying perfluoroalkyl carbon chain length and branching as well as trace levels of perfluorocarboxylic acid impurities [2, 20, 22]. The most basic surfactant derived from the perfluoroalkyl sulfonyl fluoride raw material is the corresponding sulfonate, $R_FSO_3^-$. Perfluorooctane sulfonate (PFOS) has historically been made in the largest amounts. Perfluorohexane sulfonate (PFHxS) and perfluorodecane sulfonate (PFDS) are also commercially relevant [23]. Recently, the major historic manufacturer of long-chain perfluoroalkyl sulfonyl chemistry, including PHxSF, POSF, and PDSF, ceased their production and moved to the manufacture of PBSF-based fluorinated surfactants (e.g., $C_4F_9SO_2$-R) which are growing in commercial use [24].

Using the perfluoroalkyl sulfonyl fluoride, for example, PBSF, as a basic building block, unique products are created through the sulfonyl moiety using conventional hydrocarbon reactions. Perhaps the most versatile intermediates from the ECF process are those containing the perfluoroalkyl sulfonamido functionality, $R_FSO_2N(R)$-. For example, $C_4F_9SO_2N(CH_3)CH_2CH_2OH$, *n*-methyl perfluorobutyl sulfonamido ethanol (MeFBSE). These primary alcohols can readily be functionalized into fluorinated ethoxylates, phosphates, sulfates, and (meth)acrylate monomers. Fluorinated (meth)acrylates undergo free-radical polymerizations to give oligomeric fluorinated surfactants [25–28].

In addition, perfluoroalkyl carboxylic acids (PFCAs) and their derivatives have also been synthesized using the ECF process. Typically, an alkyl carbonyl fluoride (for example $C_7H_{15}COF$) is transformed into the corresponding perfluoroalkyl carbonyl fluoride (for example $C_7F_{15}COF$). The carbonyl fluoride is then reacted to yield esters, amides, or carboxylic acid salts which are have all been commercially produced and used as surfactants [4]. The most widely known is the ammonium salt of perfluorooctanoic acid ($C_7F_{15}COOH{\cdot}NH_3$), whose major historical use has been as a processing aid in the manufacture of fluoropolymers [29].

2.2 *Telomerization: Fluorotelomers*

The free-radical addition of TFE to pentafluoroethyl iodide yields a mixture of perfluoroalkyl iodides with even-numbered fluorinated carbon chains. This is the process used to commercially manufacture the initial raw material for the "fluorotelomer"-based family of fluorinated substances (Fig. 3) [2, 17]. Telomerization may also be used to make terminal "iso-" or methyl branched and/or odd number fluorinated carbon perfluoroalkyl iodides as well [2]. The process of TFE telomerization can be manipulated by controlling the process variables, reactant ratios, catalysts, etc. to obtain the desired mixture of perfluoroalkyl iodides, which can be further purified by distillation. While perfluoroalkyl iodides can be directly hydrolyzed to perfluoroalkyl carboxylate salts [29, 30], the addition of ethylene gives a more versatile synthesis intermediate, fluorotelomer iodides. These primary alkyl iodides can be transformed to alcohols, sulfonyl chlorides, olefins, thiols, (meth)acrylates, and from these into many types of fluorinated surfactants [3] (Fig. 3). The fluorotelomer-based fluorinated surfactants range includes nonionics, anionics, cationics, amphoterics, and polymeric amphophiles.

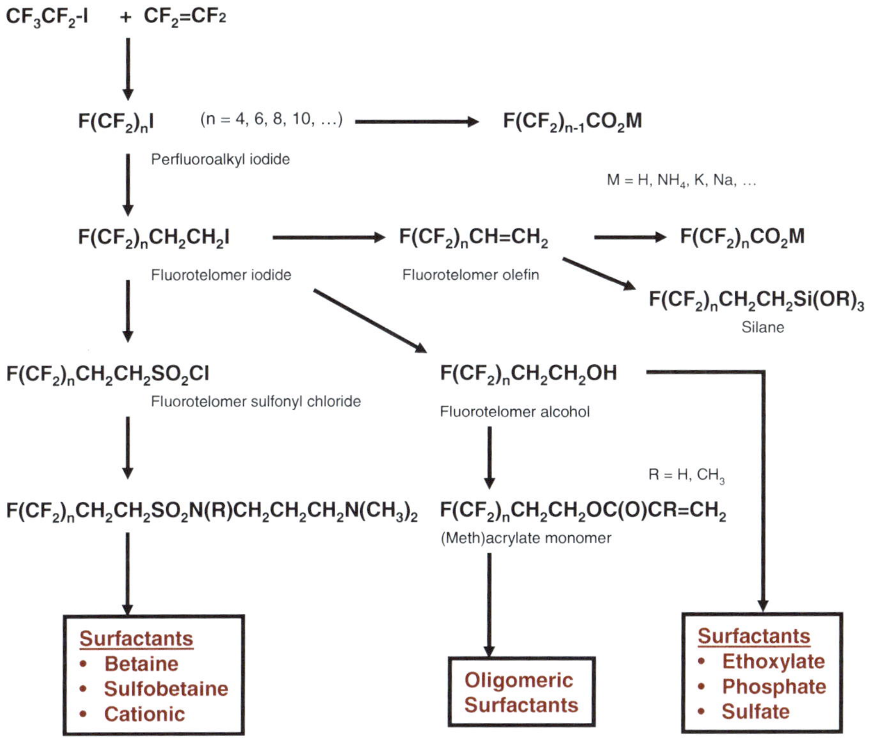

Fig. 3 Synthesis if fluorotelomer-based fluorinated surfactants

2.3 *Per- and Poly- Fluorinated Ethers*

Per- and poly- fluorinated ether-based fluorinated surfactants typically have 1, 2, or 3 perfluorinated carbon atoms separated by an ether oxygen, depending on the route to the perfluoropolyether intermediate [31] (Fig. 4). The photooxidation of TFE or HFP gives oligomers or polymers with mono- or di-acid end groups [18]. These perfluoropolyethers have random sequences of $-CF_2O-$ and either $-CF_2CF_2O-$ or $-CF(CF_3)CF_2O-$ units, from TFE or HFP, respectively. In general, the photooxidation of TFE yields mostly difunctional perfluoropolyether acid fluorides, while the photooxidation of HFP yields mostly the monofunctional perfluoropolyether acid fluoride [18]. The fluoride catalyzed oligomerization of HFPO [32], an epoxide, yields a mixture of perfluoropolyether acid fluorides, which can be converted to many types of surfactants, analogous to the fluorinated surfactants from the ECF syntheses.

Per- and poly-fluorinated ether surfactants are the newest commercially available substances in this rapidly expanding group of fluorinated surfactants [33–35]. For example, the phosphate shown in Fig. 4 is used as a grease repellent for food contact paper [36]. Per- and poly-fluorinated polyether carboxylates [37–41] are also used as processing aids in the synthesis of fluoropolymers. Per- and poly-fluorinated polyether silanes are used as surface treatments [42–45].

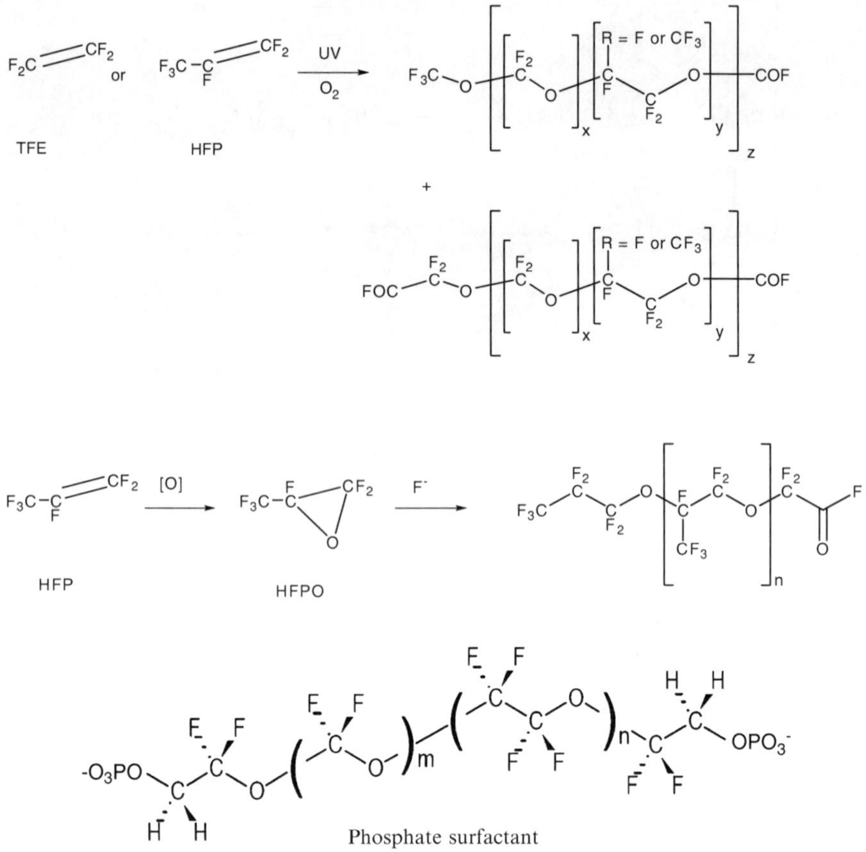

Fig. 4 Synthesis of perfluoropolyether-based intermediates and surfactants

2.4 Fluorinated Oxetanes

An alternative route to fluorinated surfactants originates from the reaction of polyfluorinated alcohols with oxetanes bearing a $–CH_2Br$ group in their side-chains to create fluorinated oxetane monomers that undergo ring-opening polymerization to give side-chain polyfluorinated polyethers (Fig. 5). Oxetane-based fluorinated surfactants are offered in many forms and functionalities, such as phosphates and ethoxylates [19, 46–48].

2.5 Spacers

Separating and joining the hydrophobic/oleophobic "tail" and the hydrophilic group of the fluorinated surfactant is the critical organic linking group, often called the "spacer" [2–6]. Perfluoroalkyl acid (PFAA) surfactants such as perfluoroalkyl

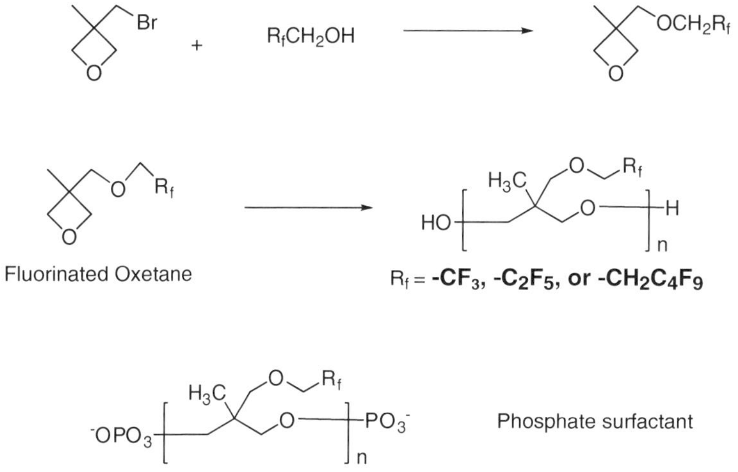

Fig. 5 Synthesis of fluorinated oxetane surfactants

carboxylates (PFCAs,) and perfluoroalkyl sulfonates (PFSAs, $F(CF_2)_nSO_3^-$), have no organic linking group between the hydrophobic/oleophobic and the hydrophilic portions of the molecule. For most fluorinated surfactants, the organic linking group provides a distance between the amphophiles, which optimizes their surface activity, intermolecular, and intramolecular interactions. The organic linking group often contains heteroatoms (nitrogen, oxygen, or sulfur) which impart a greater hydrophilicity than a mere hydrocarbon spacer. The possible combinations of (1) hydrophobic/oleophobic portion, (2) hydrophilic portion, and (3) organic linking group for fluorinated surfactants are essentially endless. A partial list of fluorinated surfactants is shown in Fig. 6 to provide an introduction to the range of the more common fluorinated surfactants and their uses.

3 Properties

The performance attributes of fluorinated surfactants are unique and distinguish them from all other types of surfactants. Fluorinated surfactants are costly and therefore are generally only used because no other alternative surfactant (e.g., hydrocarbon, silicone) can deliver the required performance. The key features of fluorinated surfactants are many. First, their surface activity in both aqueous and solvent systems is unmatched. Fluorinated surfactants can lower aqueous surface tension to less than 16 dynes/cm and function at very low concentrations (e.g., 100–500 mg/L or parts-per-million, ppm). They are effective in both basic and acidic aqueous media. In addition, fluorinated surfactants are effective in organic solvents including esters, alcohols, ethers, and solvent-based resin systems. Second, the reduced surface tension achieved by using fluorinated surfactants results in superior wetting, spreading, and leveling properties for all types of surfaces

Anionic Fluorinated Surfactants

Aqueous Film-Forming Foam (AFFF)

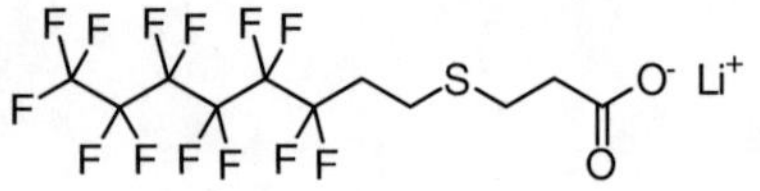

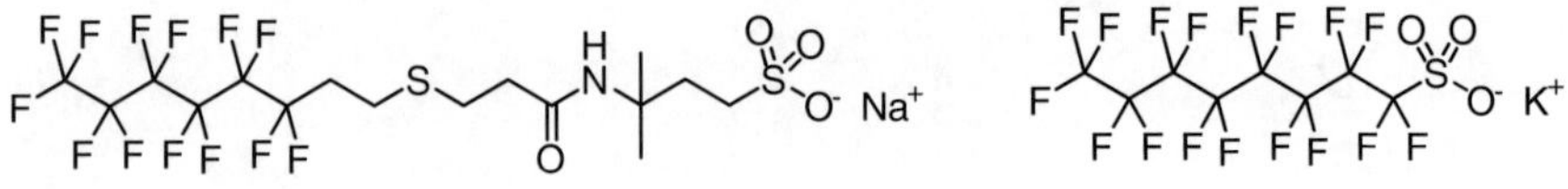

Fluoropolymer Processing Aid

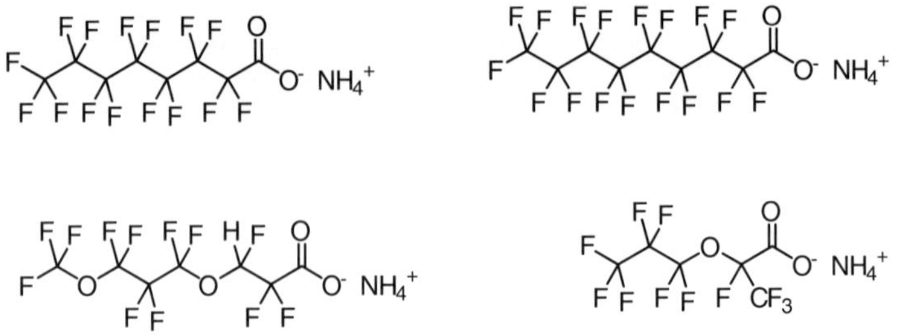

MetalPlating

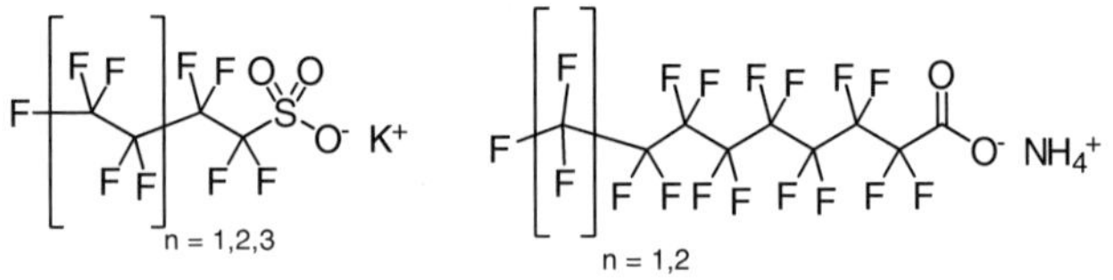

Inks

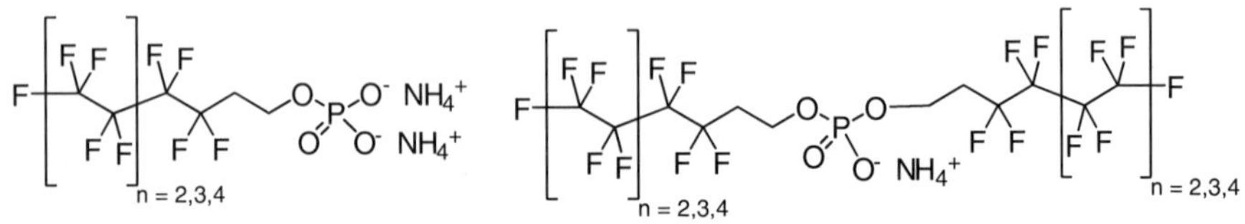

OpticalElements

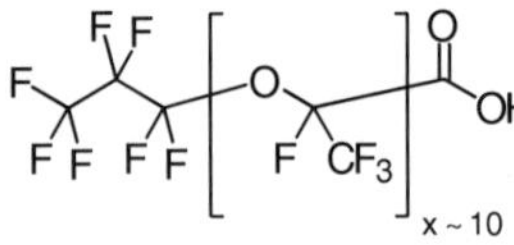

Photoresists

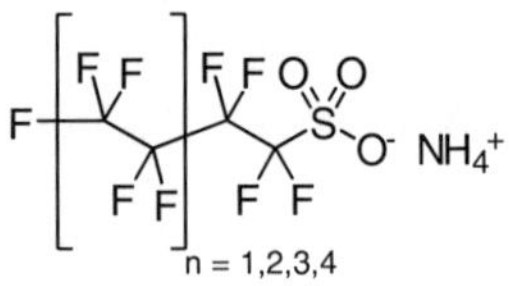

Fig. 6 (Continued)

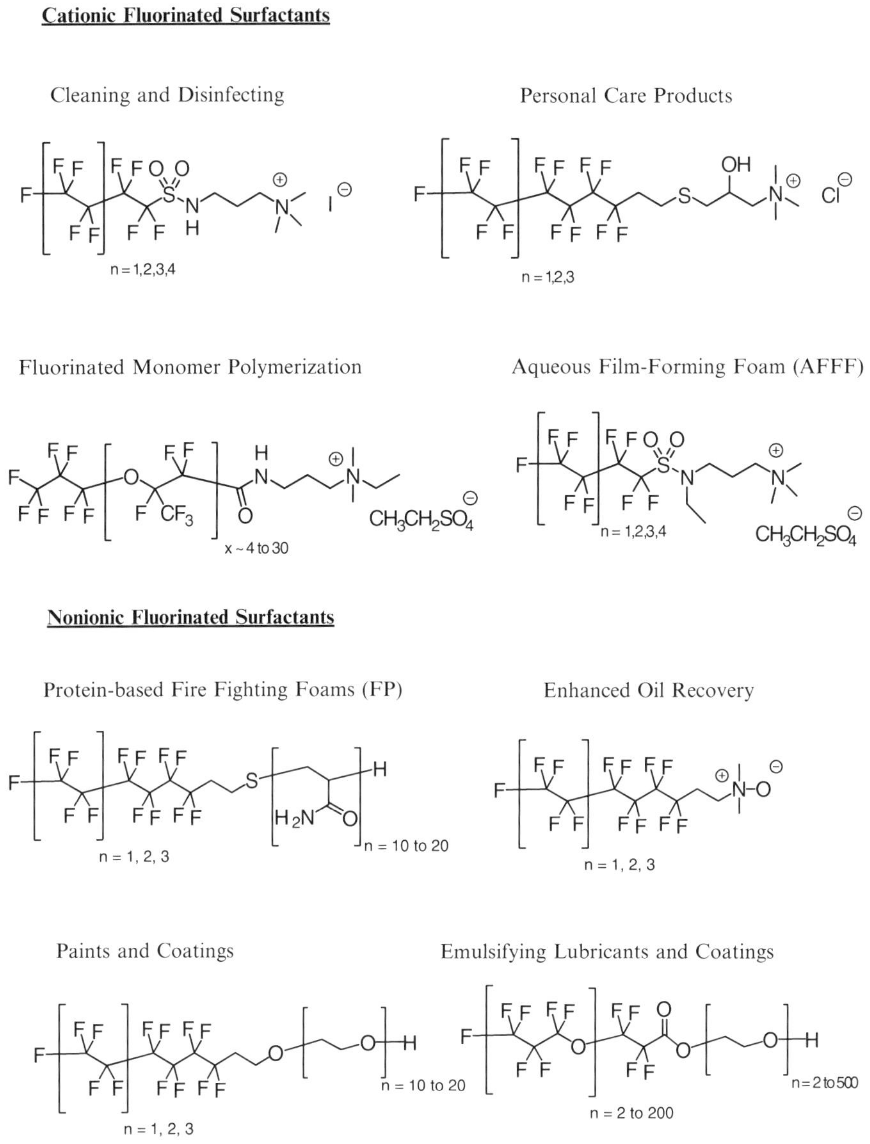

Fig. 6 (Continued)

(e.g., hard surface, wet surfaces, plastics, wood, porous surfaces and even oily metals). This package of unique performance properties, derived from the low surface energy, gives uniform film formation of coatings and eliminates pinholes and craters, even when applied to unclean surfaces. Third, fluorinated surfactants are effective emulsifiers in specialty applications where fluorinated materials are in either the dispersed or continuous phase (e.g., synthesis of fluoropolymers). Finally,

Amphoteric Fluorinated Surfactants

Aqueous Film-Forming Foam (AFFF)

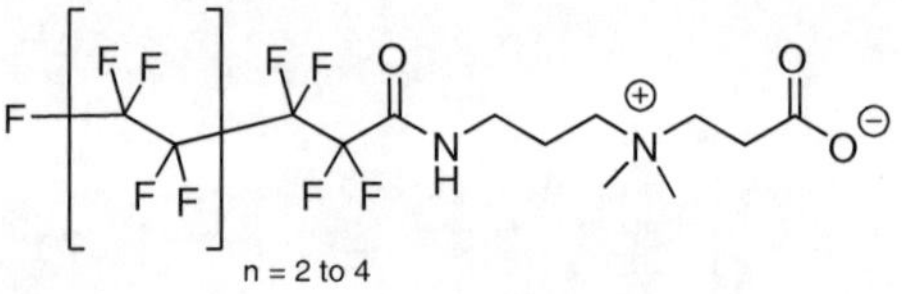

Lithographic Printing

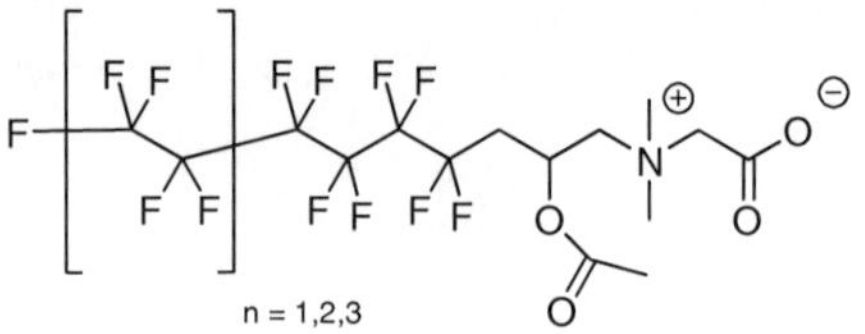

Ceramic Tile Treatment

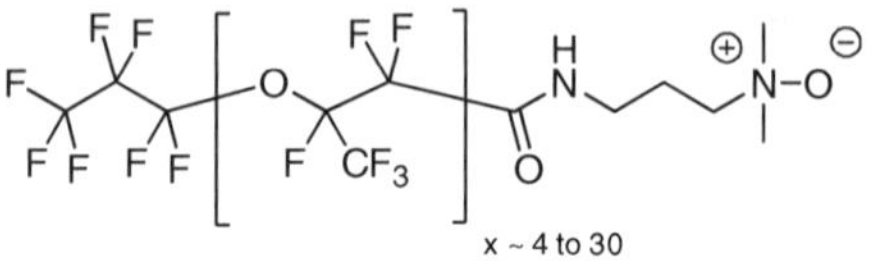

Fig. 6 Examples of fluorinated surfactants and their uses

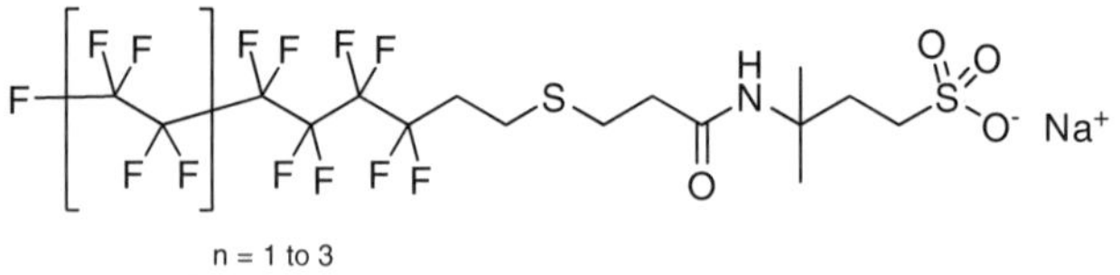

Fig. 7 Photographic film fluorinated surfactant

perfluorinated sulfonic and carboxylic acids, are extremely stable both chemically and thermally. In harsh use conditions such as hot chromic acid, concentrated sulfuric acid or hydrofluoric acid and concentrated hot alkaline solutions where other surfactants are destroyed [49], they are stable and effective in lowering surface energy or when used as foam stabilizers.

Because of their exceptionally low aqueous surface tension, fluorinated surfactants are used in applications including fire fighting foams, paints, coatings, mining, paper, electroplating, photographic emulsifiers, pressure sensitive additives, waxes, polishes, insecticides, mold release, ink jet printing, lithography, enhanced oil recovery (EOR), and emulsion polymerizations, etc. [2, 3, 6, 11] The critical micelle concentration (CMC) of a fluorinated surfactant is close to that of an ordinary hydrocarbon surfactant whose chain length is about 1.5 times longer than a fluorocarbon chain [50]. However, fluorinated surfactants with longer fluorinated hydrophobic/oleophobic chains, for example, greater than eight fluorinated carbon atoms, have reduced water solubility which limits their reduction in CMC and surface tension. Kunieda found that fluorinated surfactants have many industrial uses because they have surface tension, which is considerably lower than that of ordinary surfactants and exceptional stability against acids, alkalines, oxidizing agents, reducing reagents, and elevated temperature [51].

Recent research continues to support the earlier conclusions regarding the surface tension reduction properties of fluorinated surfactants. For example, Ngo stated that, "fluorinated surfactants are more surface-active and more hydrophobic than their corresponding hydrocarbon analogs" [52]. In Ngo's study of four cationic surfactants, an increased hydrophobic portion resulted in lower CMC and fluorinated surfactants more significantly lowered the surface tension than their corresponding hydrocarbon surfactants; see Table 1.

Sharma wrote that, "...fluorinated surfactants are about ten times more effective than silicones and 50–100 times more effective than hydrocarbon surfactants" [53]. In Sharma's study, an increased hydrophilic portion of nonionic fluorinated surfactant shifted the HLB (hydrophile–lipophile balance) and resulted in an increased CMC and an increase in minimum surface tension. The increased surface tension for the more hydrophilic nonionic fluorinated surfactant was attributed to a lower surface excess and higher surface area occupied by each surfactant molecule at the air–liquid interface, both indicating less dense packing at the interface; see Table 2.

Table 1 Surface properties of cationic surfactants

Cationic fluorinated surfactant	CMC ($\times 10^{-3}$mol/L)	γ at CMC (mN/M)
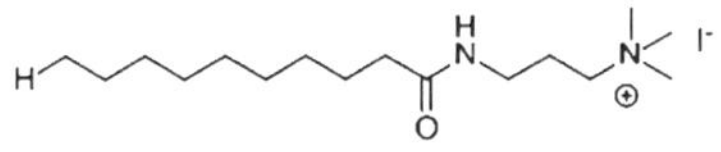	12	35
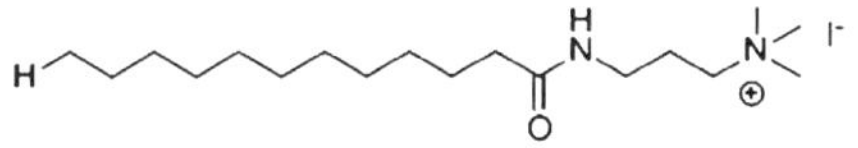	3.5	34
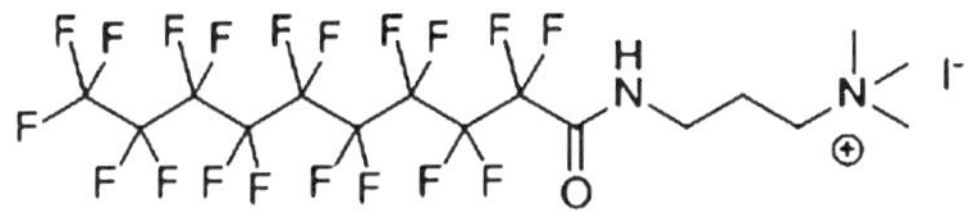	0.5	15
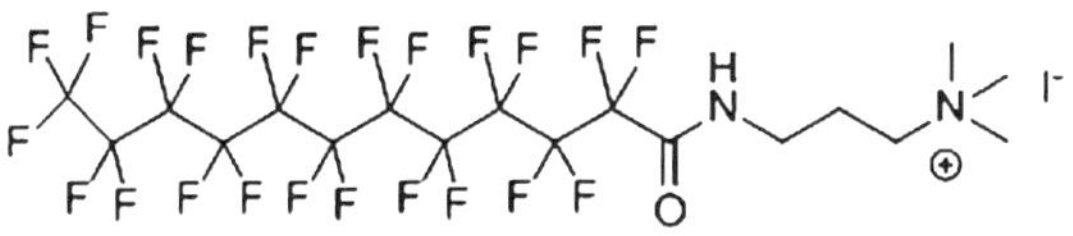	0.07	17

Surface tension measured at 45°C

Table 2 Surface properties of nonionic fluorinated surfactants

Nonionic fluorinated surfactant	CMC ($\times 10^{-6}$ mol/L)	γ at CMC (mN/M)
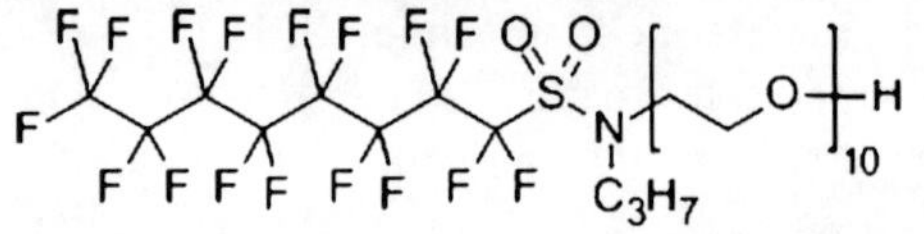	5	21
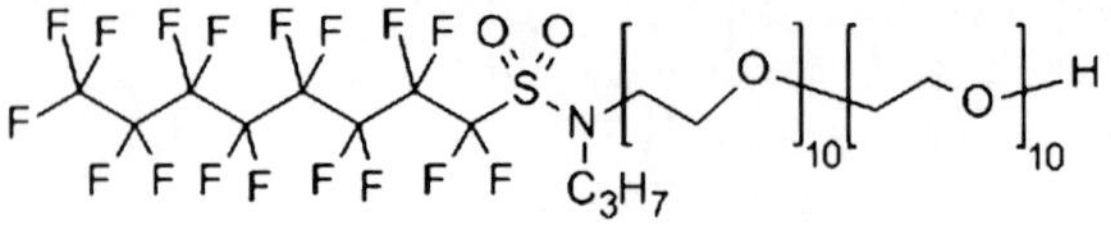	23	24.6

Surface tension measured at 25°C

Table 3 Physical properties of anionic fluorinated surfactants

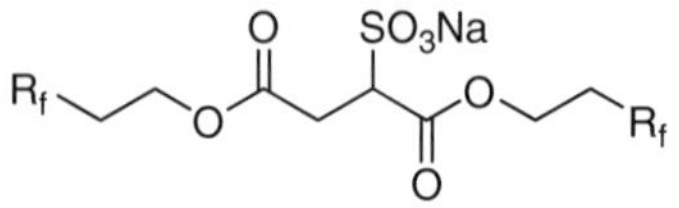

R_f	Krafft temperature	CMC ($\times 10^{-6}$ mol/L)	γ at CMC (mN/M)	Measurement temperature
$F(CF_2)_4$	<0	720	16.3	30°C
$F(CF_2)_6$	26°C	45	22.0	30°C
$F(CF_2)_8$	73°C	10	12.8	73°C

In Sagisaka's study of "The Effects Of Fluoroalkyl Chain Length ... (in) Fluorinated Anionic Surfactants," increasing chain length of the fluorinated hydrophobic/oleophobic portion of the anionic surfactant resulted in increasing Krafft temperatures and an "abrupt decrease in water solubility," see Table 3 [54]. The decreased water solubility accounted for the lowest minimum surface tension at 30°C for the C_4F_9 hydrophobe compared to the C_6F_{13} and C_8F_{17} hydrophobes.

More extensive elaboration of the properties and physical chemistry of fluorinated surfactants are discussed in recent monographs [2–6, 8, 9, 17]. The reader is referred to these and the other citations given in this chapter for more in-depth property information.

4 Commercial Uses

The first commercially manufactured fluorosurfactants were the PFAAs, PFOS, [11, 55] and PFOA [56–58] made by ECF. Their unique properties led to use in a plethora of industrial and consumer applications. Here we highlight the major commercial uses of fluorinated surfactants today. This compilation is meant to be representative, not exhaustive and does not include all known uses. Again, the reader is referred to the citations given for more information about other uses not described here.

4.1 Aqueous Film-Forming Foams

Fluorinated surfactants are particularly well suited for fire-fighting foams used to fight flammable liquid fires [7]. Fluorinated surfactants have been used for decades as critical ingredients in fire-fighting foam (aqueous film-forming foam, AFFF and Film Forming Fluoroproteinic foams, FFFP) products because of their unparalleled surface tension lowering, wetting and spreading properties [6, 59]. Historically, perfluoroalkyl sulfonates (PFSAs) such as PFOS and PFSA-based surfactant derivatives [e.g., $F(CF_2)_nSO_2N(R)R'$ where R = H, CH_3, C_2H_5, R′ = additional functional group] were the most widely used surfactants in AFFF [6, 60–63]. Alternatively, fluorinated surfactants based on fluorotelomer thiol [64], e.g., [$F(CF_2)_nCH_2CH_2SCH_2CH(OH)CH_2N^+H(CH_3)CH_2CO_2^-$] and sulfonyl [65], e.g., $F(CF_2)_nCH_2CH_2SO_2NHCH_2CH_2N^+(CH_3)_2CH_2CH_2CO_2^-$, chemistry have also been used in AFFF. Harkins established the criterion necessary to attain spontaneous spreading of two immiscible liquids [66]. Spontaneous spreading of an aqueous solution and film formation on top of the hydrocarbon surface should occur when the surface tension of the lower hydrocarbon fuel phase is greater than the sum of the surface tension of the upper aqueous phase and the interfacial tension between the aqueous upper phase and the lower hydrocarbon phase. Because the surface tension of hydrocarbon fuels and polar organic solvents is generally between 18 and 30 mN/m, only a fluorinated surfactant can provide AFFF with the required low surface tension and positive spreading coefficient that enable film formation [67]. The exceptional fire-fighting effectiveness of fire fighting foam is due to the formation and spreading of an aqueous film formed on top of lighter hydrocarbon fuels, which is accomplished by using fluorinated surfactants. To illustrate the effectiveness of fire fighting foam, a lightning strike in January 1993 caused a large explosion and fire at a Brazilian oil refinery in a tank which had 15 million liters of diesel fuel. Over 100 firefighters worked for 12 h and, by using AFFF, they prevented the fire from spreading into nearby fuel tanks and adjacent buildings [68].

4.2 Enhanced Oil Recovery

Fluorinated surfactants are effective in a variety of EOR techniques including (1) improving subterranean wetting, (2) increasing foam stability, and (3) modifying the surface properties of the reservoir formation by lowering surface tension and foaming properties to well-stimulation additives [69–72]. Both fluorotelomer [69] and ECF-based [73–77] surfactants have been and are used. EOR using a fluorinated surfactant was employed at a well in Moffat County, Colorado located in the Fort Union Sand Formation. After treatment with methanol, C_{10}–C_{12} alcohol ethoxylates, and a cationic polymeric fluorinated surfactant, the gas productivity in this well increased from 100 million cubic feet (MCF) per day to 300 MCF per day [78].

4.3 Coatings

Fluorinated surfactants uniquely provide the quintessential properties of exceptional wetting, leveling and flow control for water-based, solvent-based and high-solids organic polymer coating systems when added in amounts of just 100–500 ppm [79–84]. Fluorinated surfactants impart valuable properties to paints and coatings including anti-crater and improved surface appearance, better flow and leveling, reduced foaming, decreased block, open-time extension, oil repellency, and dirt pickup resistance [85]. They have also been widely used in inks [86]. The inclusion of fluorinated surfactants in ink jet compositions has led to better processing through modern printers and excellent image quality on porous or non-porous media [87]. Fluorinated surfactants improved surface wetting during the screen printing of carbon black inks onto Polymer Electrolyte Membrane (PEM) fuel cell electrodes [88]. In addition, fluorinated surfactants improved the cold-water swelling and internal bond strength of wood particleboard bonded with urea–formaldehyde (UF) adhesive resins due to reduced interfacial tension of the resins and improved substrate wetting [89].

4.4 Industrial and Institutional

Fluorinated surfactants are particularly useful for cleaning hard surfaces such as wood, glass, countertops, and flooring because of their ability to lower surface tension, enhance wettability, and stabilize foam. An early use that continues today is in floor polishes [90]. Cleaning compositions with cationic and nonionic fluorinated surfactants were found both to remove soil exceptionally well and to provide a protective layer which assists future cleaning of the surface by preventing or reducing the adhesion of soil subsequently deposited onto the surface [91, 92].

4.5 Electroplating and Electrowinning

Fluorinated surfactants are able to reduce the surface tension of aqueous solutions at temperatures up to 70°C which has resulted in valuable applications in the field of electroplating for both plastics such as acrylonitrile butadiene styrene (ABS) and for metals. Electroplating is mainly used to deposit Chromium. One of the challenges of this application is not only to have a surfactant stable in the presence of hot chromic acid (e.g., concentrations of 350 g/L at 70°C) but it also needs to resist decomposition during the electrolysis. Under these demanding conditions, perfluorinated surfactants such as PFOS, PFHxS, PFBS, PFOA, and PFNA are stable and maintain their activity over a longer period than a fluorotelomer-based surfactant such as 6:2 fluorotelomer sulfonate, $C_6F_{13}CH_2CH_2SO_3^-$. Under electrolytic conditions, the hydrogen atoms in the ethylene spacer of a fluorotelomer-based surfactant may be easily abstracted leading to surfactant decomposition and loss of surfactant properties. Perfluorinated acids are also used in the electrowinning of copper because they are stable and provide surface tension lowering as well as stable foam formation that aids in acid mist suppression [93].

4.6 Electronics

The foaming properties of fluorinated surfactants are widely recognized in aqueous foam systems but fluorinated surfactants are also used to stabilize foam in polar solvents such as isopropanol. Isopropanol foams are used in the electronic industry and particularly for surface preparation. The metallic surface from which greases and contaminants need to be removed before welding passes on top of an isopropanol foam maintained by the incorporation of fluorinated surfactants and a constant injection of air into the liquid. Fluorinated surfactants can stabilize a foam in a polar solvent by forming micelles with hydrophobic interiors and hydrophilic exteriors which are compatible with the polar environment. Micellar formation results in the lowering of the liquid–air interfacial tension even in a polar solvent. The electronics industry uses fluorinated surfactants in aqueous solutions for acid etching of silicon wafers as well as for the preparation of the copper-containing substrates [94].

4.7 Paper

Fluorinated surfactants have been evaluated for paper uses since the early 1960s [13, 95, 96]. Perfluorooctyl sulfonamido ethanol-based phosphates were the first substances used to provide grease repellence to food contact papers [97–99]. Fluorotelomer thiol-based phosphates and polymers followed [100–102]. Since paper fibers and phosphate-based fluorinated surfactants are both anionic, cationic bridge molecules need to be used in order to ensure the electrostatic adsorption of the surfactant onto the paper fiber. These surfactants are added to paper through the wet end press where cellulosic fibers are mixed with paper additives before entering the paper forming table of a paper machine. This treatment provides excellent coverage of the fiber with the surfactant and results in good folding resistance. An alternative treatment method involves application of a grease repellent at the size press and film press stage which consists of impregnating the formed paper sheet with a surface treatment. Fluorinated phosphate surfactants are not preferred for this mode of paper treatment. In this latter case, fluorinated polymers are used instead of surfactants. In terms of oil and water repellency, it is well recognized in the paper industry that phosphate-based fluorinated surfactants provide good oil repellency but have limited water repellency. Acrylate polymers with fluorinated side chains derived from sulfonamido alcohols and fluorotelomer alcohols are the most widely used polymers because they deliver oil, grease, and water repellence. Most recently, perfluoropolyether-based phosphates and polymers have become widely used treatments for food contact paper and paper packaging [36].

4.8 Mining

As discussed earlier, fluorinated surfactants are used in many applications because of their ability to stabilize aqueous foams and remain stable under strongly acidic and strongly basic conditions. This is the case for fire-fighting foams and EOR. In the mining industry, fluorinated surfactants are used to create stable aqueous foams for ore flotation to separate metal salts from soil and in electrowinning of metals such as copper [94].

4.9 Photographic Films

One of the challenges for the design of photographic films is the build-up of electrostatic charge [103] during film manufacturing, during transport in cameras or in photofinishing equipment. When the overcoat of the photographic film is based on gelatin and hydrocarbon surfactants, positive charges are created. The incorporation of fluorinated surfactants effectively reduces the static charging of the overcoat [104]. When friction occurs during handling, the photographic film then becomes neutral or slightly anionic which lowers the overall charge accumulation on the film and reduces potential exposure marks in the light sensitive layers of the film.

4.10 Fluoropolymer Polymerization Aid

Fluorinated surfactants have been used for decades as processing aids during aqueous emulsion polymerization synthesis of fluoropolymers such as poly (tetrafluoroethylene). The function of the fluorosurfactant is to solubilize both the fluorinated monomer(s) as well as the growing fluoropolymer. Historically, the most widely used surfactants for emulsion polymerization are the ammonium salts of perfluorooctanoic and perfluorononanoic acid [29]. Currently the fluoropolymer industry is working toward the elimination of the use of these acids, primarily through development of alternatives such as carboxylates of per- and poly-fluorinated ethers [41]. (See also Sect. 2.3, vide infra.)

4.11 Pesticide Application

Fluorinated surfactants have been used as formulation additives to aid in the delivery of pesticides and have been identified as degradation products of pesticidal active ingredients. Perfluoroalkyl phosphonic acids (PFPAs), $O = P(OH)_2C_nF_{2n+1}$, and perfluoroalkyl phosphinic acids (PFPiAs), $O = P(OH)(C_nF_{2n+1})(C_mF_{2m+1})$ are commercial surfactants manufactured and offered for a range of consumer and industrial applications including past use as inert additives in pesticide formulations [105–107]. Fluorotelomer alcohol-based phosphates have been approved for this use as well. Recently, the approval for use of these surfactants has been withdrawn [107]. The insecticide sulfluramid (*N*-ethyl perfluorooctanesulfonamide) was developed for control of ants and cockroaches and degrades in the environment to form perfluorooctanesulfonamide, $C_8F_{17}SO_2NH_2$, and PFOS. Registrations for this insecticide have been withdrawn in the United States but are still permitted in some countries.

5 Summary

The world of fluorinated surfactants is full of many useful and unique products tailored for specific end users who take advantage of their exceptional performance properties. Recently, major global manufacturers have made commitments to work toward eliminating the manufacture of "long-chain" perfluoroalkyl carboxylates and perfluoroalkyl sulfonates and substances that may break down to them in the environment [108–110]. As a result, commercial production has shifted to short chain alternatives [24] and new fluorinated moieties such as the per- and poly fluorinated ethers. Clearly, there remains a need for fluorinated surfactants in many industries to obtain the beneficial performance properties of these substances that cannot be achieved with other types of surfactants.

References

1. Kissa E (2001) Fluorinated surfactants and repellents. Surfactant Science Series 97. Marcel Dekker, New York
2. Kissa E (1994) Fluorinated surfactants: synthesis-properties-applications. Surfactant Science Series 50. Marcel Dekker, New York

3. Taylor CK (1999) Fluorinated surfactants in practice. Annu Surfact Rev 2:271–316
4. Lehmler HJ (2009) Production of fluorinated surfactants by electrochemical fluorination. Surfact Sci Ser 142:301–321
5. Lehmler HJ (2005) Synthesis of environmentally relevant fluorinated surfactants-a review. Chemosphere 58(11):1471–1496
6. Schultz MM, Barofsky DF, Field JA (2003) Fluorinated alkyl surfactants. Environ Eng Sci 20(5):487–501
7. Pabon M, Corpart JM (2002) Fluorinated surfactants: synthesis, properties, effluent treatment. J Fluorine Chem 114(2):149–156
8. Smart BE (1994) Characteristics of C-F systems. In: Banks RE, Smart BE, Tatlow JC (eds) Organofluorine chemistry. Plenum Publishing, New York, NY
9. Banks RE, Smart BE, Tatlow JC (1994) Organofluorine chemistry: principles and commercial applications. Plenum, New York
10. Simons JH (1949) Electrochemical process for the production of fluorocarbons. J Electrochem Soc 95:47–59
11. Holzapfel W (1966) Uses of fluorinated chemicals. Fette, Seifen, Anstrichmittel 68 (10):837–842
12. Grajeck EJ, Peterson WH (1959) Fluorochemicals: the new idea in textiles. Am Dyest Rep 48(13):37–39
13. Brown RE (1963) Fluorochemicals and packaging. Pap Ind 45:205–208
14. Lau C et al (2007) Perfluoroalkyl acids: a review of monitoring and toxicological findings. Tox Sci 99(2):366–394
15. Butt CM et al (2010) Levels and trends of poly- and perfluorinated compounds in the arctic environment. Sci Total Environ 408(15):2936–2965
16. Houde M et al (2006) Biological monitoring of polyfluoroalkyl substances: a review. Environ Sci Technol 40(11):3463–3473
17. Rao NS, Baker BE (1994) Textile finishes & fluorosurfactants (in Organofluorine chemistry. Principles and commercial applications). Plenum, New York, pp 321–336
18. Sianesi DG (1996) Perfluoropolyether (PFPEs) from perfluoroolefin photooxidation. (In Organofluorine Chemistry. Principles and Commerical Applications.) Plenum, New York, pp 431–462
19. Thomas RR (2006) Other types of surfactants – fluorinated surfactants. (In Chemistry and Technology of Surfactants). Blackwell, Oxford, UK, pp 227–235
20. 3M Company (1999) Fluorochemical use, distribution and release overview. US EPA Admin Record AR226-0550
21. 3M Company (1995) 3M The Leader in Electrofluorination (3M Company Technical Bulletin) St. Paul, MN
22. Arsenault G et al (2008) Some issues relating to the use of perfluorooctanesulfonate (PFOS) samples as reference standards. Chemosphere 70(4):616–625
23. 3M Company (2000) Re: Phase-out Plan for POSF-based Products. US EPA Admin Record AR226-0588
24. Ritter SK (2010) Fluorochemicals go short. Chem Eng News 88(5):12–17
25. Savu PM, Etienne SA (2001) Fluorochemical sulfonamide surfactants. EU Appl 2,001,027. 3M Innovative Properties Company, USA
26. Dams RJ et al. (2006) Compositions of monomeric surfactants. US Appl 2,006,148,671. 3M Innovative Properties Company, USA
27. Terrazas MS et al. (2006) Fluorochemical polymeric surfactants. US Appl 2,006,149,012. 3M Innovative Properties Company, USA
28. Moore GGI, Dams RJ (2008) Fluorinated surfactants and method of making the same. World Appl 2,008,045,999,200,8. 3M Innovative Properties Company, USA
29. Prevedouros K et al (2006) Sources, fate and transport of perfluorocarboxylates. Environ Sci Technol 40(1):32–44

30. Grottenmuller R et al. (2002) Process for the preparation of perfluorocarboxylic acids from perfluoroalkyl iodides and oxygen. EU Patent 1,172,350. Clariant, GmbH
31. Kasai PH (1995) Perfluoropolyethers with acid end groups: amphiphilicity and emulsification. J Appl Polym Sci 57(7):797–809
32. Millauer H, Schwertfeger W, Siegemund G (1985) Hexafluoropropene oxide – a key compound of organofluorine chemistry. Angew Chem 97(3):164–182
33. Audenaert FA, Dams RJ (2006) Fluorinated polyether copolymer surfactants. World Appl 2,006,112,2. 3M Innovative Properties Company, USA
34. Audenaert FA, Dams RJ, Tan LS (2005) Mixture of fluorinated polyethers, liquid compositions and surfactant use. US Appl 2,005,096,244. 3M Innovative Properties Company, USA
35. Dams RJ et al. (2009) Preparation of fluorinated ether compounds used to reduce surface tension, make foam, and treat surfaces. World Appl 2,009,071,5. 3M Innovative Properties Company, USA
36. Solvay-Solexis (2010) Solvera® PFPE. http://www.solvaysites.com/sites/solvayplastics/EN/specialty_polymers/Fluorinated_Fluids/Pages/Solvera_PFPE.aspx. Accessed 15 December 2010
37. Visca M et al. (2003) Aqueous fluoropolymer dispersions containing bifunctional fluorinated surfactants with reduced environmental impact. World Appl 2,003,279,9. Solvay Solexis S.p.A., Italy
38. Higuchi S et al. (2007) Aqueous polytetrafluoroethylene emulsions for fine powders and porous materials with good paste extrusion molding property. World Appl 2,007,046,345. Asahi Glass Co., Ltd., Japan
39. Hintzer K et al. (2007) Method of making fluoropolymer dispersion. US Appl 2,007,015,864. 3M Innovative Properties Company, USA
40. Ishikawa T, Tsuda N, Yamamoto Y (2009) Preparation of fluoro ether carboxylic acids useful for surfactants and aqueous dispersion used thereof. US Appl 2,009,176,942. Daikin Industries, Ltd., Japan
41. Gordon SC (2011) Toxicological evaluation of ammonium 4,8-dioxa-3 H-perfluorononanoate, a new emulsifier to replace ammonium perfluorooctanoate in fluoropolymer manufacturing. Reg Tox Pharmacol 59(1):64–80
42. Dams RJ et al. (2002) Compositions comprising fluorinated polyether silanes for rendering substrates oil and water repellent. World Appl 2,002,030,848. 3M Innovative Properties Company, USA
43. Dams RJ, Martin SJ, Flynn RM (2005) Fluorinated polyether isocyanate derived silane compositions with good oil and water repellency and stain resistance. US Appl 2,005,054,804. 3M Innovative Properties Company, USA
44. Jinks PA, Dams RJ (2009) Medicinal inhalation devices comprising multifunctional polyfluoropolyether silane applied to surface. World Appl 2,009,061,902. 3M Innovative Properties Company, USA
45. Moore GGI et al. (2003) Coating composition containing fluoro-containing polyether silane polycondensate with durable repellency to water, oil and stain. World Appl 2,003,040,209. 3M Innovative Properties Company, USA
46. Kausch CM et al (2003) Surface tension and adsorption properties of a series of bolaamphiphilic poly(fluorooxetane)s. Langmuir 19(18):7182–7187
47. Kausch CM et al (2003) Interfacial rheological properties of a series of bolaamphiphilic poly (fluorooxetane)s. Langmuir 19(18):7354–7361
48. Kausch CM et al (2002) Synthesis, characterization, and unusual surface activity of a series of novel architecture, water-dispersible poly(fluorooxetane)s. Langmuir 18(15):5933–5938
49. Dams R (1993) Fluorochemical surfactants for hostile environments. Spec Chem 13(1):4–6
50. Shinoda K, Hato M, Hayashi T (1972) The physicochemical properties of aqueous solutions of fluorinated surfactants. J Phys Chem 76(6):909–914

51. Kunieda H, Shinoda K (1976) Krafft points, critical micelle concentrations, surface tension, and solubilizing power of aqueous solutions of fluorinated surfactants. J Phys Chem 80 (22):2468–2470
52. Ngo THV et al (2010) Synthesis and comparative behaviour study of fluorocarbon and hydrocarbon cationic surfactants in aqueous media. J Fluorine Chem 131(6):704–708
53. Sharma SC, Shrestha LK, Aramaki K (2007) Interfacial properties of aqueous nonionic fluorocarbon surfactant solutions. J Dispersion Sci Technol 28(4):577–581
54. Sagisaka M et al (2001) Effects of fluoroalkyl chain length and added moles of oxyethylene on aggregate formation of branched-tail fluorinated anionic surfactants. Colloids Surf A 183–185:749–755
55. Burdon J et al. (1957) Fluorinated sulphonic acids. Part I. Perfluoromethane-,octane-, and -decanesulfonic acids and their simple derivatives. J Chem Soc:2574–2578
56. Segal L et al (1958) Oil- and water-repellent treatments for cotton with fluorochemicals. Tex Res J 28:233–241
57. Philips FJ, Segal L, Loeb L (1957) Application of fluorochemicals to cotton fabrics to obtain oil- and water-repellent surfaces. Tex Res J 27:369–378
58. Kauck EA, Diesslin AR (1951) Some properties of perfluorocarboxylic acids. Ind Eng Chem 43(10):2332–2334
59. Fiala R (1982) Use of extinguishing agents in aircraft post-crash fire-fighting/rescue. AGARD Lect Ser 123:8/1–8/27
60. Tuve RL, Jablonski EJ (1966) Perfluoro compounds as foams for hydrocarbon fires. US 3,258,423. United States Dept. of the Navy
61. Clarke M (1992) AFFFs – Fighting Aircraft Fires. Aircraft Engineer November 12–13
62. Moody CA, Field JA (2000) Perfluorinated surfactants and the environmental implications of their use in fire-fighting foams. Environ Sci Technol 34(18):3864–3870
63. Pabon M, Corpart JM (1999) Fluorinated surfactants in fire fighting foams. Actualite Chimique 7:3–9
64. Falk RA (1984) Perfluoroalkyl amphoteric compounds. EU Patent 1,152,51. Ciba-Geigy A-G, Switzerland
65. Bertocchio R, Foulletier L (1971) Beta -[2-[2-(Perfluoroalkyl)ethylsulfonylamino]ethyl] dimethylammoniopropionates. DE 2120868. Ugine Kuhlmann, Germany
66. Harkins WD, Feldman A (2009) Films. The spreading of liquids and the spreading coefficient. J Am Chem Soc 44:2665–2685
67. Cortina T (2009) Aircraft rescue and firefighting requires the best foam available: AFFF. Asia-Pacific Fire Magazine June: 25–28
68. Figueredo RCR and Sabadini FALReE (1999) Science of foams – application in fire-fighting. Quim. Nova, 22(1) 126–130
69. Murphy PM, Hewat T (2008) Fluorosurfactants in enhanced oil recovery. Open Petroleum Eng J 1:58–61
70. 3M Company (1992) Fluorad® well stimulation additives. St. Paul, MN
71. Clark HB, Pike MT, Rengel GL (1982) Water-soluble fluorochemical surfactant well stimulation additives. J Pet Technol 34(7):1565–1569
72. Clark HB (1980) Use of fluorochemical surfactants in nonaqueous stimulation fluids. J Pet Technol 32(10):1695–1697
73. Baran JR et al. (2009) Methods for treating hydrocarbon-bearing formations with fluorinated polymer compositions. World Appl 2,009,085,904. 3M Innovative Properties Company, USA
74. Baran JR, Terrazas MS, Dams RJ (2009) Fluorinated anionic surfactants and fluoropolymers in well treatment fluids for petroleum recovery. World Appl 2,009,085,899,200,9. 3M Innovative Properties Company, USA
75. Dams RJ, Baran JR, Wu YK (2010) Method of contacting hydrocarbon-bearing formations with fluorinated phosphate and phosphonate compositions. World Appl 2,010,080,353,201,0. 3M Innovative Properties Company, USA

76. Dams RJ (2010) Method of contacting hydrocarbon-bearing formations with fluorinated ether compositions. World Appl 2,010,080,473,201,0. 3M Innovative Properties Company, USA
77. Moore GGI, Dams RJ, Baran JR (2009) Preparation of fluorinated polymers for use in treating hydrocarbon-bearing formations. World Appl 2,009,085,936. 3M Innovative Properties Company, USA
78. Pope GA et al. (2009) Use of fluorocarbon surfactants to improve the productivity of gas and gas condensate wells. US Appl 2,009,062,155. Board of Regents, The University of Texas System, USA; 3M Innovative Properties Company, USA
79. Linert JG, Chasman JN (1993) Applications of fluorochemical surfactants in high-solids and water-borne coatings. Proc. Water-Borne, Higher-Solids, Powder Coat Symp:462–476
80. De Witte J, Piessens G, Dams R (1995) Fluorochemical intermediates, surfactants and their use in coatings. Surf Coat Int 78(2):63–64
81. Witte JD, Piessens G, Dams R (1994) Fluorochemical intermediates, surfactants and their use in coatings. Fluorine Coat., Int. Conf., Paper 17
82. Linert JG, Chasman JN (1993) Applications of fluorochemical surfactants in high-solids and water-borne coatings. Am Paint Coatings J. December 20:37–43
83. Moore GGI et al. (2007) Compositions of fluorochemical surfactants and their use in coatings. US Appl 2,007,049,646. 3M Innovative Properties Co., USA
84. Pike MT (1972) Fluorochemical surfactants for nonaqueous systems. Paint Varn Prod 62(3):27–32
85. Meng J (2006) Novel applications for fluorosurfactants in low VOC coatings. FutureCoat Proc:051/1–051/6
86. Dams R (1995) Fluorochemical surfactants for inks. Pitture Vernici Eur 71(13):30–2
87. Ma Z et al. (2001) Color ink jet ink composition with fluorosurfactant. US6,436,180,200,1. Hewlett-Packard Co., USA
88. Wills RGA, Watt-Smith MJ, Walsh FC (2009) The use of fluorocarbon surfactants to improve the manufacture of PEM fuel cell electrodes. Fuel Cells 9(2):148–156
89. Mansouri HR et al (2007) Fluorinated polyether additives to improve the performance of urea-formaldehyde adhesives for wood panels. J Appl Polym Sci 106(3):1683–1688
90. Bultman DA, Pike MT (1981) The use of fluorochemical surfactants in floor polish. Chem Times Trends 4(1):41–4
91. Dams RJ, Terrazas MS (2008) Aqueous compositions of fluorinated surfactants and coating/ cleaning use on surfaces and reducing surface tension of liquids. World Appl 2,008,147,796. 3M Innovative Properties Company, USA
92. Leach MJ, Niwata Y (1998) Cleaning composition comprising fluoro-surfactants. EU Patent 866115. Unilever PLC, UK
93. Anich AT, Sierakowski MJ (1995) Mist suppressant for solvent extraction metal electrowinning. World Appl 9530783. Minnesota Mining and Manufacturing Co., USA
94. Savu PM, Lamanna WM, Parent MJ (2005) Fluorinated sulfonamide surfactants for aqueous cleaning solutions. US Appl 2,005,197,273. 3M Innovative Properties Company, USA
95. Rengel GL, Perrault EW, Rambosek GM (1966) Fluorochemical sizing agent for wet-end addition. Tappi 49(2):76A–79A
96. Olson CC et al. (2005) Fluorochemical sizing. Sizing Pap. (3rd edn):151–178
97. Schwartz C (1981) Oil resistance utilizing fluorochemicals. Sizing, Short Course Notes Tech. Assoc. Pulp Pap Ind:71–5
98. Chad RM (1982) Use of fluorochemicals in reprographic paper grades. Coat Conf Proc:47–55
99. Allison MC (1984) Application of fluorochemicals in paper surface coatings. Pap Technol Ind 25(1):13–16
100. Giatti R (1996) New fluorochemicals for greaseproof papers. Pap Technol 37(9):49–54
101. Deisenroth E et al (1998) The designing of a new grease repellent fluorochemical for the paper industry. Surf Coat Int 81(9):440–447

102. Liu Y, Jho C (2004) Sizing with fluorochemicals: internal application. J Pulp Pap Sci 30 (2):45–49
103. Orem MW, Andrews RR (2004) Fluorinated surfactants in overcoat compositions and elements containing same. US Appl 2,004,137,385. Eastman Kodak Company, USA
104. Dams RJ, Arren DHC, Brandon M (1996) Fluoropolymers and fluorochemical surface active agents for improving the antistatic behavior of materials and light-sensitive material having improved antistatic behavior. EU Appl 1,994,071,819,96. Minnesota Mining and Manufacturing Co., USA
105. D'Eon JC et al (2009) Perfluorinated phosphonic acids in Canadian surface waters and wastewater treatment plant effluent: discovery of a new class of perfluorinated acids. Environ Toxicol Chem 28(10):2101–2107
106. U.S. Environmental Protection Agency (2006) Meeting with Registrant on Perfluorinated Chemical (Masurf-FS-780, Fluowet PL 80-B, and Fluowet PL-80) Use as Inerts. Letter from the Office of Pollution Prevention and Toxics to the Office of Pesticide Programs. U.S. EPA Public Docket AR226-3804
107. U.S. Federal Register (2006) Inert Ingredient; Revocation of the Tolerance Exemption for Mono- and Bis-(1 H, 1 H, 2 H, 2 H-perfluoroalkyl) Phosphates Where the Alkyl Group is Even Numbered and in the C6-C12 Range. 71(153):45408–45411
108. OECD (2010) OECD Portal on Perfluorinated Chemicals. http://www.oecd.org/site/0,3407,en_21571361_44787844_1_1_1_1_1,00.html. Accessed 15 December 2010
109. U.S. Environmental Protection Agency (2009). Long-Chain Perfluorinated Chemicals (LCPFCs) Action Plan. http://www.epa.gov/opptintr/existingchemicals/pubs/actionplans/pfcs.html. Accessed 15 December 2010
110. U.S. Environmental Protection Agency (2006) 2010/2015 PFOA Stewardship Program. EPA-HQ-2003-0012-1071. http://www.epa.gov/opptintr/pfoa/pubs/pfoastewardship.htm. Accessed 15 December 2010

Perfluorinated Compounds: Occurrence and Uses in Products

Stefan Posner

Abstract Perfluorinated compounds are a chemical family of all organic compounds consisting of a carbon backbone fully surrounded by fluorine and represent a large and complex group of organic substances with unique characteristics. They are used in several industrial branches, but they also occur in a large range of consumer products. Because of their extraordinary properties such as chemically inert, non-wetting, very slippery, nontoxic, nonstick, highly fire resistant, very high-temperature ratings, highly weather resistant, they are applied in fluoropolymer-coated cookware, sports clothing, extreme weather-resistant military uniforms, food handling equipment, medical equipment, motor oil additives, fire fighting foams, paint and ink as well as water-repellent products. Currently, the knowledge of the exact chemical compositions in articles and preparations of perfluorinated compounds is very limited. Since the exact composition of perfluorinated compounds in consumer products is mostly confidential, a range of analytical studies concerning the content of perfluorinated compounds in consumer products have been carried out over the past years with the intention to better understand the intentional and residual content and release of fluorinated substances from consumer products and their impact to health and the environment.

Keywords Consumer products • Perfluorinated carboxylic acids • Perfluorinated compounds • Telomer alcohols

Contents

S. Posner
Swerea IVF, Box 104, 431-22 Mölndal, Sweden
e-mail: stefan.posner@swerea.se

T.P. Knepper and F.T. Lange (eds.), *Polyfluorinated Chemicals and Transformation Products*, Hdb Env Chem (2012) 17: 25–39,
DOI 10.1007/978-3-642-21872-9_2, © Springer-Verlag Berlin Heidelberg 2012

Abbreviations

FTOH	Fluorotelomer alcohols
FTS	Fluorotelomer sulfonates
PFCA	Perfluoroalkyl carboxylic acid/Perfluoroalkyl carboxylate
PFCs	Perfluoroalkyl compounds
PFNA	Perfluorononanoic acid
PFOA	Perfluorooctanoic acid
PFOS	Perfluorooctane sulfonic acid
PFS	Perfluorinated sulfonates
POP	Persistent organic pollutant
PTFE	Polytetrafluoroethylene
UNEP	United Nations Environmental Programme

1 Introduction

Perfluoroalkyl compounds (PFCs) do not occur naturally. They have been manufactured for 50 years and represent a large and complex group of organic substances with unique characteristics that are extremely versatile and used in a variety of industrial and household applications. Presently the knowledge of the exact chemical compositions in articles and preparations of perfluorinated compounds is very limited.

Recent years of research have substantially improved our knowledge of this wide range of compounds and their uses, but still there is a lot to explore concerning their uses, their intrisic properties and occurrence in the environment.

The main characteristics of polyfluorinated compounds are the replacement of most hydrogens by fluorine in the aliphatic chain structure. Some of these organic fluorine compounds are known as perfluorinated, which means that all hydrogens have been replaced with fluorine with a large variety of chemical forms and structures. Because of the diversity of fluoro organic substances, it is important to understand the developed chemical terminology.

2 The Family of PFCs

PFCs are a chemical family of all organic compounds consisting of a carbon backbone fully surrounded by fluorine, which makes them impervious to heat, acid or other forces that typically break down chemical compounds. They are used in several industrial branches, but they also occur in a large range of consumer products. Because of their extraordinary properties (chemically inert, non-wetting, very slippery, nontoxic, nonstick, highly fire resistant, very high temperature ratings, highly weather resistant, etc.), they are applied in fluoropolymer-coated cookware, sports clothing, extreme weather-resistant military uniforms, food handling equipment, medical equipment, motor oil additives, fire fighting foams, paint and ink as well as water-repellent products.

Fluorotelomers are a range of chemicals with similar fluoride carbon backbones connected to a $-CH_2-CH_2-$ chain and different functional heads. They are industrially produced by applying a telomerization process, coupling tetrafluoro-ethene, which leads to straight-chained products with an even number of carbon atoms. Fluorotelomers are probably the most commonly used perfluorinated substances in products. The hydroxyl group as functional group will give fluorotelomer alcohols (FTOH). They are used to treat paper to improve its moisture and oil barrier properties. FTOHs are also used in waterproof outdoor clothing and in waterproofing agents for textiles. Fluorotelomer alcohols are manufactured as a raw material used in the synthesis of fluorotelomer-based surfactants and polymeric products.

The manufacture of FTOHs usually results in a mixture containing 6–12 fluorinated carbon congeners, the 8:2 FTOH being the dominant one. Release of the volatile FTOHs may occur all along the supply chain from production, application into consumer use and disposal. They have the potential to form stable perfluorinated carboxylates (PFCAs) such as perfluorooctaoic acid (PFOA) and perfluorononaoic acid (PFNA) which are shown in Fig. 1.

The general chemical structure of perfluorinated sulphonates (PFS) contains a perfluorinated carbon chain connected to a sulphonate group. In addition to this, fluorotelomer sulphonates (FTSs) contain two carbon atoms adjacent to the functional group that are not fluorinated. FTSs are used among other fluorotelomers in fire fighting foam for their film-forming properties and the ability to decrease fuel absorption. These foams are especially useful against major fires, e.g., chemical fires (Stockholm Convention on POPs Review Committee 2009). The quantities in the foams are low, but the foams are released directly into the environment.

FTS is also used as a component in more complex structures (e.g., in water proofing agents) and as a substitute for perfluorooctane sulphonate (PFOS).

Fluorinated surfactants are used in very low levels in a large number of cleaning products, e.g., polish, waxes, all-purpose cleaners, window cleaners, etc. Their use is widespread and directly released into wastewater.

PFCA is another important PFC group. The main use of perfluorooctanoate (PFOA) is as a process aid in the manufacture of various fluoropolymers, such as polytetrafluoroethylene (PTFE). These polymers are among other things, used to coat cookware intended for stovetop cooking and baking.

The substances PFOS and PFOA are part of a group of old-generation PFCs which will be used to a lesser extent in the future because of their potential hazards. These hazards have resulted and will result in a number of international legislative bans worldwide. New generations of PFCs are developed continuously and applied in industrial amounts already.

Polyfluorinated sulphonamides are considered the most important PFCs because of their intentional industrial production and global distribution. PFOS and related

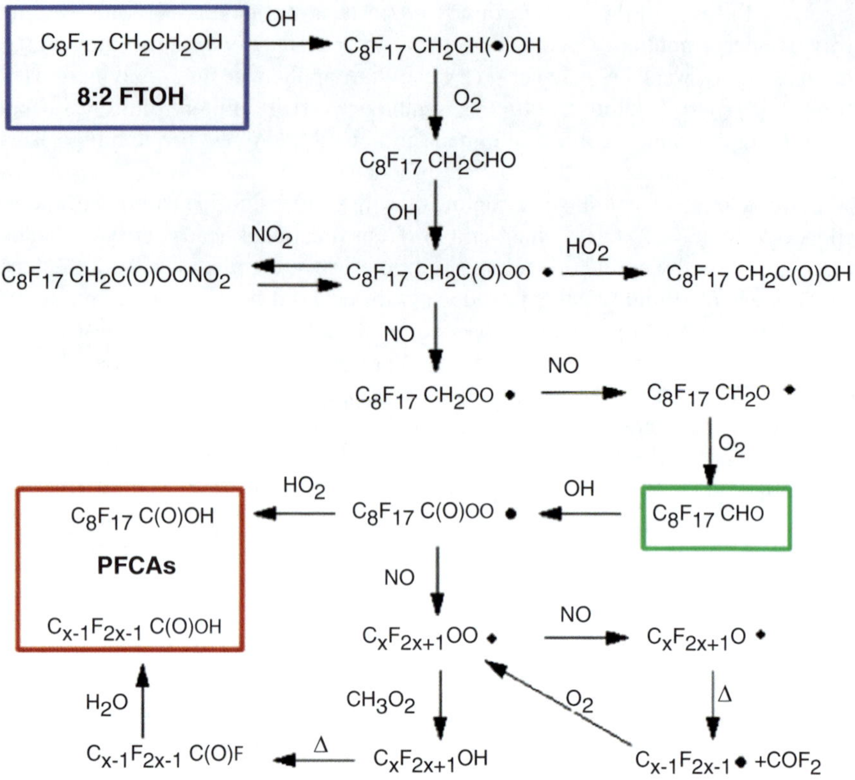

Fig. 1 Simplified mechanism for the atmospheric degradation of 8:2 FTOH into perfluorocarboxylic acids (*red box*), Wallington et al. [1]

substances are well-known degradation products from substituted sulphonamides that are used commercially for numerous applications. However, because of their potential toxicity, extreme persistence and accumulation potential of their degradation product, PFOS has resulted in prohibition on new uses or import by chemical regulatory authorities worldwide based on international restrictions by the United Nations Environmental Programme (UNEP) Stockholm convention, where PFOS is going to be classified as a POP (Persistent Organic Pollutant).

Other PFCs, such as perfluoroalkylsulphonic acid derivatives (e.g., PFOSF), are probably used as paper additives/coatings to prevent oil from soaking through or staining the paper.

3 Physical Properties of PFC

Surface energy is the most critical parameter in the action of PFCs. Due to their extraordinary properties (chemically inert, non-wetting, very slippery, nontoxic, nonstick, highly fire resistant, very high-temperature ratings, highly weather resistant, etc.), they are applied in fluoropolymer-coated cookware, sports clothing, extreme weather-resistant military uniforms, food handling equipment, medical equipment, motor oil additives, fire fighting foams, paint and ink as well as water-repellent products.

Therefore, it is essential to define these surface properties in order to achieve the appropriate surface protective properties or otherwise the purpose of the surface treatment is lost.

PFCs can therefore be used to provide water repellence, stain resistance and soil release properties to a treated surface which is related to the physical properties of these fluorinated materials. The critical surface tension is the determining physical parameter why fluorinated chemicals can repel both water and oil substances [2, 3].

The critical surface energy γ_c of the CF_3 and CF_2 groups are much lower compared with the surface energy of the corresponding hydrocarbons (CH_3 and CH_2), which is described in Table 1.

One of the fundamental laws of physics states that every system strives for a minimal surface energy. Therefore, when a PFC is coated on a textile substrate and exposed to water with its surface tension of 72 mN/m or oily substances with surface tensions of 20 mN/m and more, they will not spread on the textile surface. The consumer can observe this phenomenon as "water and oil repellence".

The spreading of a liquid on a surface is measured via contact angles and demonstrates well when a fabric is being wetted or not (Fig. 2).

As can be seen from the formula for spreading $S = \gamma_c - (\gamma_L + \gamma_{cL})$, it is observed that if the surface energy of the substrate is lowered sufficiently, the liquid will not be able to wet the surface.

Practice shows that it is not sufficient to have only terminal CF_3 groups in a fluorinated chemical. Optimum reduction of the surface energy γ_c is achieved with perfluorinated chains with a sufficient chain length to obtain a large enough density of fluorinated carbons on the surface.

Table 1 Surface energies for characteristic polymer backbone structures

Surface	Liquids	Surface energy: γ_c (mN/m)	Surface tension: γ_L (mN/m)
$-CF_3$		6	
$-CF_2H$		15	
$-CF_2-$		18	
$-CH_3$		22	
$-CH_2-$		31	
$-CH_2CHCl-$		39	
Polyester		42	
Polyamide		46	
Cotton		44	
	Water		72
	n-Octane		22
	Olive oil		32

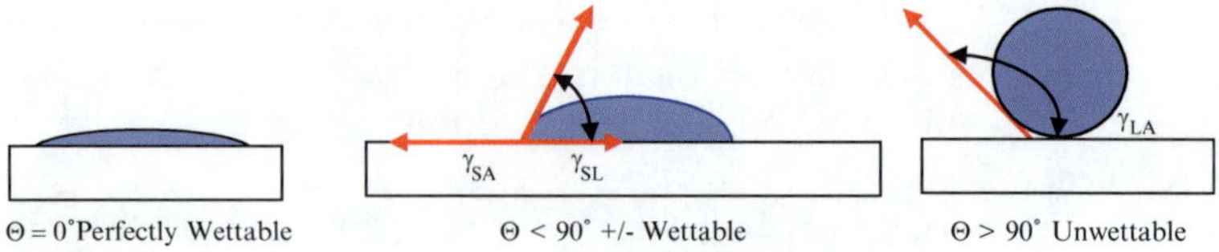

Fig. 2 Contact angle versus wettability of a substrate surface. When angle Θ is $> 90°$, liquid will not wet the surface; when angle Θ is $<90°$, liquid will wet surface partially; when angle $\Theta = 0°$, complete spreading & wetting of the surface by the liquid. Spreading occurs only if $S > 0$. Spreading coefficient: $S = \gamma_{SV} - (\gamma_{LV} + \gamma_{SL})$ S = solid, L = liquid, A = air. where γ_{SA} = surface energy of the substrate (e.g., polymer surfaces), γ_{LA} = surface tension of the liquid and γ_{SL} = interfacial tension

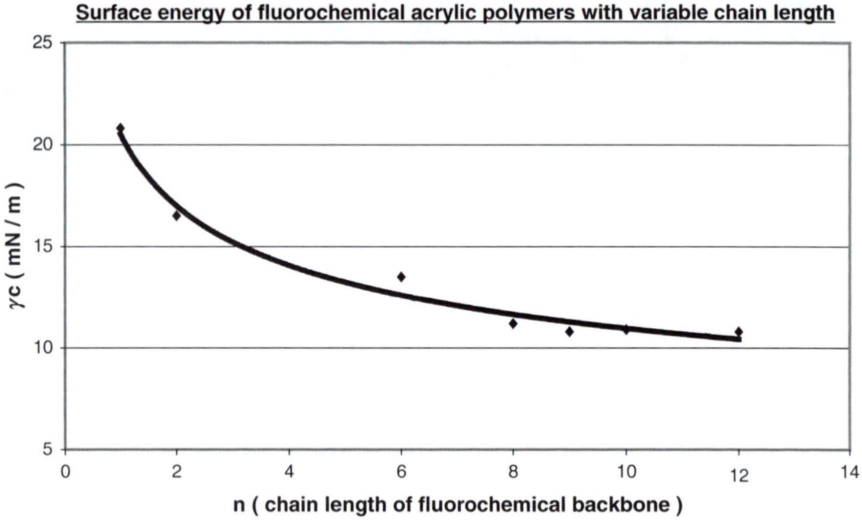

Fig. 3 Surface energy versus the number of carbons in PFC backbone structure, where 4 on the *X*-axis means four perfluorinated carbons, etc.

This has been demonstrated in the literature [2, 3] on fluorochemicals that there is a relationship of the chain length of the perfluorinated chains that is related to the critical surface energy of the surface as described in Fig. 3.

On the basis of the already explained surface energy properties of fluorinated chemicals, it is understood that for instance a non-fluorine surface treatment, such as silicones on treated polymers, can provide rather good water repellency, but no oil repellency due to the fact that the oil has lower surface energy than that the silicone layer has. The surface energy obtained with a silicone surface cannot be lower than 22 mN/m, which is comparable to the surface tension of hydrocarbon, oily substances. This means that fluorotelomers are not always possible to replace with a non-fluorine surface treatment if oil or soil repellence is required.

4 Historic Emissions of PFCs

PFCs have been manufactured for more than 50 years where the substances PFOS and PFOA are part of a group of old-generation PFCs which will be used to a lesser extent in the future due to their potential hazards. These hazards have resulted and will result in a number of international legislative bans worldwide.

In the European Union, in the REACH regulation, PFOS and its precursors are the only EU-regulated PFC substance and PFOA is assessed concerning its intrinsic properties which resulted in a classification as toxic (T; R48/23), carcinogenic (Carc Cat2, R45) and a reproductive toxicant (Repr Cat2, R61). Presently, no national or European regulation of the use of PFOA exists, but it may be the case in the future.

United States Environmental Protection Agency (US EPA) banned PFOS and its precursors since 2001 and has a voluntary agreement with the fluoropolymer industry for emissions of PFOA and its homologues. In this global stewardship programme on perfluorooctanoic acid (PFOA) and related chemicals, the industry commit to reducing PFOA and related chemicals by 95% no later than 2010, and to work towards total elimination of PFOA from emissions and in products no later than 2015.

Australia is developing definitions of and limit uses to non-dispersive applications, similar to restrictions in EPA consent orders with companies.

Due to this pressure from the international society, new generations of PFCs are developed continuously and applied in industrial amounts already.

Due to identified hazards for PFOS and PFOA in the context of their long history of production, their historical global use and emissions of perfluorinated compounds in the last 50 years are of major environmental importance.

Prevedouros and his research group published a model in 2006, where cumulative global emissions between 2005 and 2050 have been predicted to be at least 80% lower than the estimated cumulative emissions between 1950 and 2004.

Composing an initial global-scale mass balance model to evaluate the identified direct emissions of PFCs from manufacture and using this model could account for observed concentrations of PFCs in the environment (Prevedouros et al. [4]) (Table 2).

According to the modelling work of Prevedouros and his team for the long-term (1950–2050) global fate of PFOA, they identified direct emissions of PFOA from manufacture and use that could account for observed concentrations of PFOA in the environment [4].

Table 3 lists both direct and indirect PFOA emission sources to the global environment. This table also presents estimated minimum and maximum projected cumulative emissions between 2005 and 2050 for each source together with the contribution of each source to the total PFOA emissions for the time period.

Direct sources with up to 6,900 tons emitted perfluorinated carboxylic acids (PFCA) were representing the vast majority of PFCAs emitted to the global environment compared to indirect sources which contributed up to 350 tons.

Table 2 Global historical PFCA production and emissions summary, taken from [4]

Environmental input source	Historical time period (years)	Estimated total global historical PFCA emissions (metric tons)	Estimated total global production (metric tons)
Direct PFCA sources			
(1) PFCA manufacture			
PFO/APFO	1951–2004	400–700	3,600–5,700
PFN/APFN	1975–2004	70–200	800–2,300
Total manufactured		470–900	4,400–8,000
(2) Industrial and consumer uses			
Fluoropolymer manufacture (APFO)	1951–2004	2,000–4,000	
Fluoropolymer dispersion processing (APFO)	1951–2004	200–300	
Fluoropolymer manufacture (APFN)	1975–2004	400–1,400	
Fluoropolymer processing (APFN)	1975–2004	10–20	
Aqueous fire fighting foams (AFFF)	1965–1974	50–100	
Consumer and industrial products	1960–2000	40–200	
Total direct		3,200–6,900	
Indirect PFCA sources:			
(1) POSF-based products			
PFCA residual impurities	1960–2002	20–130	
POSF-based precursor degradation	1960–2002	1–30	
POSF-based AFFF	1970–2002	3–30	
(2) Fluorotelomer-based products			
PFCA residual impurities	1974–2004	0.3–30	
Fluorotelomer-based precursor degradation	1974–2004	6–130	
Fluorotelomer-based AFFF	1975–2004	<1	
Total indirect		30–350	
Total source emissions (direct+indirect)		3,200–7,300	

[a] Low and high estimated values as well as the period of use/production for each source are based upon publicly available information cited in the text

Note: *APFO* Ammonium perfluorooctanoate; *APFN* Ammonium perfluorononanoate; *AFFF* Aqueous fire-fighting foam; *POSF* Perfluorooctylsulphonyl fluoride

The contribution of indirect sources is expected to decrease both in absolute numbers and relative to direct sources within the year 2050 [5].

5 PFCs in Articles and Their Exposure

The potential health risks associated with perfluorocarboxylic acids (PFCAs) have promoted intensive research on the sources, transport, transformation and distribution of these chemicals and their precursors in environmental media, as well as research related to ways to reduce the health risks. Despite the significant progress that has been made so far, researchers are yet to reach a consensus on what are the

Table 3 Estimated historical between 1950 and 2004 and estimated emissions between 2005 and 2050 of PFOA EMISSIONS [5]

PFOA emission source	1950–2004 min–max (metric tons)	% of total PFOA emission (average)	2005–2050 min–max (metric tons)	% of total PFOA emissions (average)
Direct SOURCES				
FP manufacturing (APFO)	2,060–4,090	72.3%	410–815	86.0%
APFO manufacturing	370–590	11.8%	20–40	4.2%
FP dispersion (APFO)	215–340	6.8%	45–75	8.7%
AFFF-ECF	50–100	1.8%	0	0%
FP manufacturing (APFN)	3–10	0.1%	<1–2	0.1%
Consumer and industrial products	2–10	0.1%	0	0%
APFN manufacturing	1–2	0%	<1	0%
PVDF (APFN)	<1	0%	<1	0%
Direct sources	2,700–5,140	92.9%	475–932	99.0%
Indirect sources				
POSF raw material degradation	4–585	5.0%	0	0%
POSF impurities	14–110	1.2%	0	0%
POSF-AFFFs	2–23	0.2%	0	0%
FT raw material degradation	3–60	0.6%	1–14	0.8%
FT impurities	<1–17	0.1%	<1–4	0.2%
Indirect sources	23–795	7.1%	1–18	1.0%
Direct and indirect sources	2,723–5,935	100.0%	476–950	100.0%

[a]*AFFF* Aqueous film-forming foams (also aqueous fire-fighting foams); *APFN* Ammonium perfluorononoate; *APFO* Ammonium perfluorooctanoate; *ECF* Electrochemical fluorination, a process used to produce fluorinated chemicals; *FP* Fluoropolymer; *FT* Fluorotelomer; *POSF* Perfluorooctanesulphonyl fluoride; *PVDF* Polyvinylidene fluoride

most important routes by which the general population is exposed to these chemicals. In particular, there are different opinions on whether PFCA-containing products are significant contributors to the total exposure.

A risk characterization from the potential exposure to PFOA in consumer articles has been published [6]. The authors investigated potential human exposure to PFOA in a wide variety of consumer articles, including treated textiles, and concluded that the trace levels of PFOA present would not be expected to cause adverse human health effects, not contributing to quantifiable levels of PFOA in human blood. The authors noted that PFOA was present in a number of consumer articles, which were not treated with fluorinated products. This may result from the presence of PFOA contamination globally [7].

A more recent study by Fromme et al. [8] used the data from indoor measurements in Canada and Norway and estimated that, for the general population in Western countries, the inhalation of house dust contributed only 0.6% to the mean PFOA daily intake and 8.2% to the high PFOA daily intake.

Tittlemier et al. [9] identified treated carpeting as the second most important source of exposure for PFOA after ingestion of food. A study by Trudel et al. found that the consumption of contaminated food is the most important pathway causing exposure to PFOA, followed by ingestion of dust and inhalation of air in low- and intermediate-exposure scenarios. Their study also found that direct, product-related exposure is dominant in high-exposure scenarios, in which consumers regularly use PFC-containing products, such as impregnation sprays, or have treated carpets in their homes.

Trudel and his co-workers also observed that product-related exposure tends to be more important for PFOA than for PFOS, most likely because PFOS is no longer used in consumer products. It is, therefore, apparent that the paucity of indoor source and exposure data contributes to the significant uncertainty and differences of opinion about the most prevalent exposure routes for these compounds.

The fact that elevated levels of PFCAs have been detected in house dust in Japan, Canada and the United States strongly suggests the presence of indoor sources. It is well known that fluorotelomer and fluoropolymer products are sources of PFCAs and that PFCAs may exist in fluorotelomer products as unwanted by-products and in fluoropolymer products as residuals. Because a broad range of commercial products either contain or are treated with fluorotelomer and fluoropolymer products, they can be potential sources of PFCAs. Given that products are often used in close proximity to humans, it is hypothesized that they can contribute to human exposure to PFCAs either directly by dermal contact and hand-to-mouth transfer or indirectly through inhalation of suspended particles from treated carpet and other interior surfaces.

There have been several studies on the PFCA content in products, but most of them report a single compound namely PFOA. In 2005, Washburn and his colleagues reported the PFOA content in 14 article groups based on theoretical calculations and analytical measurements.

Of these groups, pre-treated carpeting and carpeting treated with carpet-care solution had the highest PFOA loadings: 0.2–0.6 and 0.2–2 mg of PFOA per kg of article, respectively. Studies by other researchers reported PFOA content in non-stick cookware, food contact paper, thread sealant tape and dental floss. Data for other PFCAs in commercial products are rather scarce. One study by Sinclair et al. reported the C5 to C12 PFCA content in three brands of popcorn packaging paper.

Friends of the Earth, Norway, published a report about PFCs in impregnation fluids, covering PFCAs as well. Thirteen commercial products were analysed for a variety of PFCs. Seven of the investigated products contained PFOA, varying between 45 and 692 ng/mL [10].

In 2009 the US EPA analysed 116 commercial articles purchased from retail outlets in the United States between March 2008 and May 2008 to determine the extractable content of C5 to C12 PFCAs [11]. Of the 13 article categories, the US EPA concluded that the most important PFCA sources were carpets, stone/tile/wood sealants, textiles and textile care products (Table 4).

Table 4 Sample breakdowns of PFCAs by article category

Category	Samples	Maximum conc. of PFCAs
Pre-treated carpeting	9	292 ng/g fibre
Commercial carpet-care liquids	9	8,860 ng/g liquid
Household carpet/fabric-care liquids and foams	12	1,710 ng/g liquid
Treated apparel	16	235 ng/g product
Treated home textile and upholstery	14	437 ng/g product
Treated non-woven medical garments	5	334 ng/g product
Treated floor waxes and stone/wood sealants	11	939 ng/g product
Treated food contact paper	5	15.3 ng/g paper
Membranes for apparel	10	12.8 ng/g product
Thread sealant tapes and pastes	10	40.6 ng/g product
Non-stick cookware	14	0.00985 ng/cm^2 coated surface
Dental floss and plaque removers	8	5.81 ng/g liquid
Miscellaneous[1]	7	82.6 ng/g product

[1]Includes four carare products, two boatare products, one deck cleaner and one dry sack for outdoor use.

For most article categories, the PFCA content in a small number of the analysed samples were significantly higher than in the rest of samples.

PTFE is, for example, used to coat cookware intended for stovetop cooking and baking. Other PFCs, such as fluorotelomer and perfluoroalkylsulphonic acid derivatives (e.g., PFOSF), are or have been used to treat paper to improve its moisture and oil barrier properties. In particular, papers used in contact with high-fat content foods may be treated with fluorotelomer or perfluoroalkyl-sulphonyl-based paper additives/coatings to prevent oil stains or oil soak through the paper. Typically, these paper coatings/additives are phosphate esters or acrylic polymers containing polyfluoroalkyl functionality [12].

Larsen et al. [13] detected small amounts of PFOA (up to 140 ppb) in extracts of PTFE resins, obtained after applying pressure and increased temperatures to the material. Subsequent studies of cookware, coated with PTFE dispersions, have shown no detectable levels of PFOA extractable from cookware under normal use conditions [14]. A later study by the Norwegian Institute of Public Health (2007) went on further by detailing about these findings. In a worst case scenario the new study showed that an adult human would be exposed to 66 ng PFOA/kg bw, when drinking 100 ml water cooked in a PTFE coated pan. It was concluded that, even at an assumption of 100% uptake of PFOA, these extremely low levels will not be an essential intake route for humans. According to Horowitz, 98% of the PFOA intake is contributed to by food [15].

Begley et al. [12] analysed several consumer products for PFOA and concluded that fluorotelomer-based paper coating/additive formulations before application onto paper have the highest PFOA content, but during normal application rates this amount of PFOA will be diluted by about 300 times on the final paper product (Table 5). Therefore, the PFOA content in finished paper should be in the few 100 mg/kg range, which is consistent with the data shown in Table 5.

Table 5 Summary of PFOA analysis in product, [12]

Consumer products	Concentration of PFOA (μg/kg)
PTFE cookware	4–75
Dental floss (PTFE based)	3
Dental tape (PTFE based)	4
PTFE film/sealant tape	1,800
FEP (fluoro-ethylene-propene copolymer) tubing	nd
Popcorn bags	6–290
French fry box	nd
Paper plates (soak-proof shield)	nd
Hamburger and sandwich wrapper	nd
Perfluoro paper coatings (not applied)	88,000–160,000

nd non-detects

Table 6 Released amounts (ng) and concentrations (μg/m^2) of PFOA and FTOH from non-stick frying pans of four different brands [16]

			PFOA		6:2 FTOH		8:2 FTOH	
Brand	Surface temperature (°C)	Area (cm^2)	ng	μg/m^2	ng	g/cm^2	ng	μg/m^2
1	180	640	12	19	16	25	73	114
2	229	477	32	67	97	204	298	625
3	190	670	192	287	36	54	28	42
4	205	659	40	61	<10	<15	40	61
Stainless steel	230	670	<5	<7	<10	<15	<10	<15

The residue content of PFOA in PTFE products is directly related to the processing temperatures used to make the products. Cookware and dental products use a high-temperature sintering process that should volatilize PFOA, while production of PTFE film used as, for example, sealant tape does not use that sintering process. Begley et al. [12] conclude that fluoropolymer food-contact materials do not appear to be a significant source of human exposure to PFCs (e.g., PFOA). In particular, the coated cookwares tested did not appear to be a significant source of PFOA. Furthermore, an extreme heating test (abusive) of the cookware did not appear to increase the residual amount of PFOA in the cookware. Additional PFOA did not appear to be formed during the normal use or misuse of these products.

This result were in contrast to the results of a more recent study of Sinclair et al. [16], where gas-phase release of PFOA, 6:2 FTOH and 8:2 FTOH was measured from heating non-stick frying pans and microwave popcorn bags. Gas-phase PFOA was measured in all four non-stick frying pan brands. PFOA was reported to vaporize at 189°C and decompose at temperatures higher than 234°C.

The authors suggest that residual PFOA is released from the PTFE coating to the gas phase under the normal cooking temperatures. Gas-phase concentration of PFOA varied depending on the frying pan brand, which suggests that the sintering conditions (temperature, pressure and duration) used in the coating of fluoropolymers may have an influence on the release of PFOA. In addition, PFOA was detected in water boiled for 10 min in three brands of non-stick frying pans (Table 6).

Table 7 Released amount (ng) of PFOA of fluorotelomer alcohols at making popcorn in a bag of three different brands [16]

Brand	PFOA	6:2 FTOH	8:2 FTOH
1	16	223	258
2	17	<20	<20
3	<2.5	<20	<20

In the same study, PFOA was found in the vapours produced by microwave heating of pre-packed popcorn bags. Furthermore, milligram quantities of both PFCAs and FTOHs were calculated to coat the entire surface of the package [16]. The authors noted that they were not able to explain the origin of the FTOHs from the cookware, because FTOHs are not used to manufacture cookware, and no plausible way for FTOH to be formed from PTFE is known (Table 7).

Eventually PFCAs can be present in consumer products treated with fluorinated compounds due to intentional application, in the form of an unintended residues, or due to degradation of precursor compounds such as FTOHs. It is not always possible to distinguish between these cases, since recipes of technical applications are mostly confidential or the actual composition of the used mixture of active compounds confidential. Products intended for contact with food seem to contain small PFCA amounts, but since almost all available data origin from authors related to fluoropolymer-manufacturing companies, the interpretation of these data should be done carefully. The same is the case for research on metabolism of PFCAs in organisms. Data from independent research groups are needed in order to confirm these potentially prejudiced findings.

References

1. Wallington TJ, Hurley MD, Xia J, Wuebbles DJ, Sillman S, Ito A, Penner JE, Ellis DA, Martin J, Mabury SA, Nielsen OJ, Andersen MPS (2006) Formation of C7F15COOH (PFOA) and other perfluorocarboxylic acids during the atmospheric oxidation of 8:2 fluorotelomer alcohol. Environ Sci Technol 40:924–930
2. Colbert JF, May MA (1983) TAPPI Seminar notes – Nonwovens, 113–118
3. Corpart JM, Dessaint A (1997) Melliand Textileberichte 9:625–629
4. Prevedouros K, Cousins IT, Buck RC, Korzeniowski SH (2006) Sources, fate and transport of perfluorocarboxylates. Environ Sci Technol 40:32–44
5. Armitage J, Cousins IT, Buck RC, Prevedouros K, Russell MH, MacLeod M, Korzeniowski SH (2006) Modeling global-scale fate and transport of perfluorooctanoate emitted from direct sources. Environ Sci Technol 40:6969–6975
6. Washburn ST, Bingmann TS, Braithwaite SK, Buck RC, Buxton LW, Clewell HJ, Haroun LA, Kester JE, Rickard RW, Shipp AM (2005) Exposure Assessment and Risk Characterization for Perfluorooctanoate (PFO) in Selected Consumer Articles. ENVIRON International Corporation and DuPont
7. Yamashita N, Kannan K, Taniyasu S, Horii Y, Okazawa T, Petrick G, Gamo T (2004) Analysis of perfluorinated acids at parts-per-quadrillion levels in seawater using liquid chromatography-tandem mass spectrometry. Environ Sci Technol 38:5522–5528

8. Fromme H, Tittlemier SA, Völkel W, Wilhelm M, Twardella D (2009) Perfluorinated compounds – exposure assessment for the general population in western countries. Int J Hyg Environ Health 212(3):239–270
9. Tittlemier SA, Pepper K, Seymour C, Moisey J, Bronson R, Cao XL, Dabeka RW (2007) Dietary exposure of Canadians to perfluorinated carboxylates and perfluorooctane sulfonate via consumption of meat, fish, fast foods, and food items prepared in their packaging. J Agr Food Chem 55:3203–3210
10. Naturvernforbund (2007) Fluorerade miljögifter i impregneringsmedel. Rapport 9019009–2
11. US EPA (2009), Perfluorocarboxylic Acid Content in 116 Articles of Commerce, EPA/600/R-09/033
12. Begley TH, White K, Honigfort P, Twaroski ML, Neches R, Walker RA (2005) Perfluorochemicals: potential sources of and migration from food packaging. Food Addit Contam 22: 1023–1031
13. Larsen BS, Kaiser MA, Botelho M, Wooler GR, Buxton LW (2005) Comparison of pressurized solvent and reflux extraction methods for the determination of perfluorooctanoic acid in polytetrafluoroethylene polymers using LC-MS-MS. Analyst 130:59–62
14. Powley CR, Michalczyk MJ, Kaiser MA, Buxton LW (2005) Determination of perfluorooctanoic acid (PFOA) extractable from the surface of commercial cookware under simulated cooking conditions by LC/MS/MS. Analyst 130:1299–1302
15. Horowitz L (2007) Swiss Federal institute of Technology Zurich, Assessment of consumer exposure to perfluorinated compounds using the SceBRA-method
16. Sinclair E, Kim SK, Akinleye HB, Kannan K (2007) Quantitation of gas-phase perfluoroalkyl surfactants and fluorotelomer alcohols released from nonstick cookware and microwave popcorn bags. Environ Sci Technol 41:1180–1185
17. Dinglasan MJA, Ye Y, Edwards EA, Mabury SA (2004) Fluorotelomer alcohol biodegradation yields poly- and perfluorinated acids. Environ Sci Technol 38:2857–2864
18. Dinglasan-Panlilio MJA, Mabury SA (2006) Significant residual fluorinated alcohols present in various fluorinated materials. Environ Sci Technol 40:1447–1453
19. Annotated outline for a guidance document on perfluorooctane sulfonate alternatives (PFOS), UNEP/POPS/POPRC.5/INF/10, August 2009
20. European Community Regulation 1907/2006, REACH – Registration, Evaluation, Authorisation and Restriction of Chemicals
21. Herzke D, Schlabach M, Mariussen E, Uggerud H, and Heimstad E (2007) A literature study on selected chemical substances. SFT. TA 2238/2007
22. Jensen AA, Poulsen PB FORCE Technology. Bossi R, DMU (2008) Survey and environmental/health assessment of fluorinated substances in impregnated consumer products and impregnating agents. Danish Ministry of the Environment, Survey of Chemical Substances in Consumer Products, No. 99
23. Kubwabo C, Stewart B, Zhu J, Marro L (2005) Occurrence of perfluorosulfonates and other perfluorochemicals in dust from selected homes in the city of Ottawa. J Environ Monit 7:1074–1078
24. Martin JW, Smithwick MM, Braune BM, Hoekstra PF, Muir DCG, Mabury SA (2004) Identification of long-chain perfluorinated acids in biota from the Canadian Arctic. Environ Sci Technol 38:373–380
25. Mawn MP, Mckay RG, Ryan TW, Szostek B, Powley CR, Buck RC (2005) Determination of extractable perfluorooctanoic acid (PFOA) in water, sweat simulant, saliva simulant, and methanol from textile and carpet samples by LC/MS/MS. Analyst 130:670–678
26. Moriwaki H, Takatah Y, Arakawa R (2003) Concentrations of perfluorooctane sulfonate (PFOS) and perfluorooctanoic acid (PFOA) in vacuum cleaner dust collected in Japanese homes. J Environ Monit 5:753–757
27. Nasjonalt folkehelseinstitutt (2007) Helserisikovurdering for SFT av Teflon – Polytetrafluoretylen (PTFE) CASnr 9002-84-0

28. OECD (2006) Risk Management Series No. 21: Preliminary lists of PFOS, PFAS, PFOA and related compounds, and chemicals that may degrade to PFCA. ENV/JM/MONO(2006)15 21
29. OECD (2006) Survey on production and use of PFOS, PFAS, PFOA, PFCA, their related substances and products/mixtures containing these substances. Environment, Health and Safety Publications, Series on Risk Management, No. 22; env/jm/mono(2006)36
30. OECD. Lists of PFOS, PFAS, PFOA, PFCA, related compounds and chemicals that may degrade to PFCA. Environment Directorate, Joint meeting of the chemicals committee and the working party on chemicals, pesticides and biotechnology, ENV/JM/MONO(2006)15, 21-Aug-2007, OECD. Found at http://appli1.oecd.org/olis/2006doc.nsf/linkto/env-jm-mono(2006)15
31. Poulsen PB, Jensen AA, and Wallström E (2005) More environmentally friendly alternatives to PFOS-compounds and PFOA. Danish Ministry of the Environment. Environmental project No. 10132005
32. Rao NS, Baker BE (1994) Textile finishes and fluorosurfactants. In: Banks RE, Smart BE, Tatlow JC (eds) Orgaofluorine chemistry. Plenum, New York
33. SFT (2004) Bruken av PerFluorAkylStoffer (PFAS) i produkter i Norge.Materialstrøm-sanalyse. ISBN 82-7655-487-3
34. SFT (2006) Kartlegging av perfluoralkylstoffer (PFAS) i utvalgte tekstiler. ISBN 82-7655-285-4
35. SFT (2007) Konsekvensvurderinger av forslag til forbud mot miljøgifter i Forbrukerprodukter
36. Strynar MJ, Lindstrom AB (2008) Perfluorinated compounds in house dust from Ohio and North Carolina. Environ Sci Technol 42:3751–3756
37. Trudel D, Horowitz L, Wormuth M, Scheringer M, Cousins IT, Hungerbühler K (2008) Estimating consumer exposure to PFOS and PFOA. Risk Analysis 28:251–269
38. US EPA (2004) Preliminary risk assessment of the developmental toxicity associated with exposure to PFOA and its salts. Office of Pollution Preventiona and toxics, Risk Assessment
39. Wania F (2007) A global mass balance analysis of the source of perfluorocarboxylic acids in the Arctic Ocean. Environ Sci Technol 41:4529–4535
40. Yamada T, Taylor PH, Buck RC, Kaiser MA, Giraud RH (2005) Thermal degradation of fluorotelomer treated articles and related materials. Chemosphere 61:974–984

Mass Spectrometric Approaches to Reveal Biotransformation Products from Per- and Polyfluorinated Chemicals

Tobias Frömel and Thomas P. Knepper

Abstract In the past years, elucidation of transformation products of per- and polyfluorinated chemicals (PFC) has been a task frequently approached by analytical chemists. PCF, such as perfluorooctane sulfonate (PFOS) and perfluorooctanoic acid (PFOA) are persistent and thus, the analytical quest to detect transformation products has failed so far. Their prominence as contaminants is mainly due to their extreme persistence, which is linked to their perfluoroalkyl chain length. Molecules that are less heavily fluorinated can show very complex metabolic behavior, as is the case for fluorotelomer alcohols. These compounds are degraded via different but simultaneous pathways, which produce different stable metabolites. Biotransformation processes of PFC may occur when these compounds enter the environment, and thus known and unknown PFC may be generated in these compartments. Therefore, it is essential to determine metabolic pathways of such compounds in order to entirely understand their fate in the environment. This chapter summarizes methodological approaches and instrumental setups which have been implemented in biotransformation studies of PFC and focuses on mass spectrometric methods and the separation techniques coupled to the mass spectrometer (MS). Valuable MS approaches that have not been frequently used in studies on PFC are presented as well. Since compounds carrying C–F bonds exhibit unique properties, these will be initially presented to address the meaning of these properties both for analytical tasks and for the setup of biotransformation experiments.

Keywords CID • Fragmentation • Mass spectrometry • Per- and polyfluorinated chemicals (PFC)

T. Frömel • T.P. Knepper (✉)
University of Applied Sciences Fresenius, Institute for Analytical Research, Limburger Str. 2, 65510 Idstein, Germany
e-mail: knepper@hs-fresenius.de

T.P. Knepper and F.T. Lange (eds.), *Polyfluorinated Chemicals and Transformation Products*, Hdb Env Chem (2012) 17: 41–61,
DOI 10.1007/978-3-642-21872-9_3, © Springer-Verlag Berlin Heidelberg 2012

Contents

Abbreviations

APCI	Atmospheric pressure chemical ionization
API	Atmospheric pressure ionization
APPI	Atmospheric pressure photoionization
BPC	Base-peak chromatogram
CAD	Collision(ally)-activated dissociation
CID	Collision-induced dissociation
CRM	Charge residue model
EI	Electron impact (ionization)
ESI	Electrospray ionization
FTEO	Fluorotelomer ethoxylate
FTOH	Fluorotelomer alcohol
FTS	Fluorotelomer sulfonates
GC	Gas chromatography
HPLC	High-pressure (performance) liquid chromatography
IEM	Ion evaporation model
LC	Liquid chromatography
LIT	Linear ion trap
MRM	Multiple reaction monitoring
MS	Mass spectrometer/spectrometry/spectrometric
OECD	Organization for economic co-operation and development
PFC	Per- and polyfluorinated compounds
PFCA	Perfluorocarboxylic acid/perfluorocarboxylate
PFOA	Perfluorooctanoic acid
PFOS	Perfluorooctane sulfonate
PFSA	Perfluoroalkyl sulfonic acids
POSF	Perfluorooctanesulfonyl fluoride

Q1/q2/Q3	Quadrupoles in triple quadrupole instruments (q2 represents the collision cell)
QqQ	Triple quadrupole instrument
RP	Reversed phase
SIM	Single ion monitoring
SPE	Solid phase extraction
ToF	Time-of-flight
WWTP	Wastewater treatment plant
XIC	Extracted ion chromatogram

1 Introduction

In the past years, elucidation of transformation products of per- and polyfluorinated compounds (PFC) has been a task frequently approached by analytical chemists. It has been estimated that biotransformation contributes to approximately 0.1–5% with respect to perfluorocarboxylic acid (PFCA) historical global emissions [1]. For perfluorooctanesulfonyl fluoride (POSF)-based compounds such as perfluorooctane sulfonic acid (PFOS), biotransformation products probably affect environmental burden marginally, although no distinct estimations have been made so far [2].

Nonetheless, biotransformation processes of PFC may occur when these compounds enter the environment, and thus known and unknown PFC may be generated in these compartments. Therefore, it is essential to determine metabolic pathways of such compounds in order to entirely understand their fate in the environment. This is especially true for fluorinated polymeric materials, which have only been the focus of two scientific articles so far [3, 4], although these compounds represent a large field of application of fluorinated compounds [5].

This article summarizes methodological approaches and instrumental setups which have been implemented in biotransformation studies of PFC and focuses on mass spectrometric methods (MS, will be used for mass spectrometer, mass spectrometry and mass spectrometric) and the separation techniques coupled to the MS. Valuable MS approaches that have not been frequently used in studies on PFC are presented as well. Since compounds carrying C–F bonds exhibit unique properties, these will be initially presented to address the meaning of these properties both for analytical tasks and for the setup of biotransformation experiments.

2 Properties of Fluorine and Fluorinated Organic Compounds and the Implications for Analytical Purposes and Environmental Behavior

Fluorine is a special element within the periodic table, which results in unique and often valuable properties of fluorinated organic substances. When carrying out a biodegradation study with PFC, some of the characteristics of these substances should be kept in mind, as they may complicate gathering of reliable data, whereas others may be even helpful for the analytical chemist.

Fluorine has an atomic number of 9 and a relative atomic weight of 18.9984 u. This negative mass defect leads to substantially lower monoisotopic masses of highly fluorinated compounds than the respective nominal mass. For instance, the *m*/*z* ratio of the perfluorooctanoate anion is 412.9664. Other organic compounds usually have monoisotopic masses higher than the respective nominal mass, since most other elements have a positive mass defect. This difference can be taken advantage of by high-resolution MS.

In contrast to most other elements, fluorine is monoisotopic. Thus, fluoroorganic compounds do not exhibit characteristic isotopic patterns in MS, which is one of the disadvantageous properties of fluorine for the analytical chemist, especially the mass spectrometrist. In contrast, other organohalogens, such as organochlorines and bromines offer very pronounced isotopic patterns, which can be determined by means of MS.

Fluorine has a very small van der Waals radius of 147 pm [6] and, although very difficult to measure, a covalent radius of approximately 60 pm [7–9]. Associated with that, it has the highest electronegativity in the whole periodic system of 3.98 on Pauling's scale [10], which inevitably causes every bond A–F to have considerably ionic character, unless A is oxygen, nitrogen, or fluorine itself [7]. The C–F bond is thus better described as $C^{\delta+}$–$F^{\delta-}$.

As a result of these rather ionic interactions, the C–F bond is considered the strongest single bond in organic chemistry with a bond enthalpy of 481 kJ/mol in CH_3F, which is substantially higher than that of other bonds [11]. This pronounced bond strength is reflected in the notorious environmental and chemical stability of PFC. Another consequence of its low van der Waals radius is a very low electronic polarizability, which causes London forces and surface energies of fluorinated molecules to be very low [12] and may represent a reason for the unique partitioning characteristics of highly fluorinated molecules. They are both hydrophobic and lipophobic/oleophobic [13, 14] and, depending on the functional groups attached to the fluorinated carbon chain, have low aqueous solubility. For instance, the aqueous solubility of 8:2-fluorotelomer alcohol (8:2-FTOH) is approximately two orders of magnitude lower than its non-fluorinated counterpart 1-decanol [15]. Furthermore, 8:2-FTOH is rather liquid, whereas 1-decanol is solid, which also implies very low intermolecular forces between 8:2-FTOH molecules. Despite the low intermolecular forces, PFC tend to show distinct partitioning onto HPLC parts or environmental solids such as soil [16] or activated sludge [17] or any material used to conduct the study, for example vessels and tubes, that is, surfaces [18]. This effect may be ascribed to ionic and non-ionic interactions. As a consequence of this property, sterile controls should always be carried out simultaneously in order to differentiate between biotransformation processes and sorption.

Another effect of the low aqueous solubility of some of the compounds is volatilization, if the compounds exhibit high vapor pressure and low aqueous solubility at the same time, such as FTOH and other fluorotelomer-based biotransformation products. These compounds need to be taken care of by special instrumentation of the biotransformation experiment setup.

3 Instrumental Setup

3.1 Biodegradation Setup

Although general protocols on biotransformation/biodegradation experiments are supplied, for example, by the Organization for Economic Co-operation and Development (OECD), most scientific groups use non-standardized protocols for their investigations, which results in a great variety of parameters, like microorganisms, vessels, sampling and analysis.

Microorganisms usually originate from wastewater treatment plant (WWTP)-activated sludge [19–23], WWTP effluent [24–27] or soil [3, 4, 21, 28, 29]. Rarely, these experiments are carried out using sediment and groundwater organisms [30], mixed liquor [23, 31] or pure bacterial cultures [32]. Whereas most experiments are performed using the unaltered matrix (e.g., soil) or slightly modified matrix (e.g., wetted soil), it is also possible to grow the microorganisms in a separate vessel and transfer them into the actual vessel, which may be filled with a defined mineral medium and the test compound, for example, [30].

As for the vessels, polymer and glass tubes or bottles are routinely used, although polymer materials are considered to exhibit lower tendencies to cause adsorption of PFC. The abovementioned high sorption tendency of PFC requires the conduction of simultaneous sterilized control experiments, which are usually carried out adding biocides such as sodium azide (NaN_3) or mercuric chloride ($HgCl_2$), which are also recommended by the OECD [33]. An example is shown in Fig. 1, where adsorption of

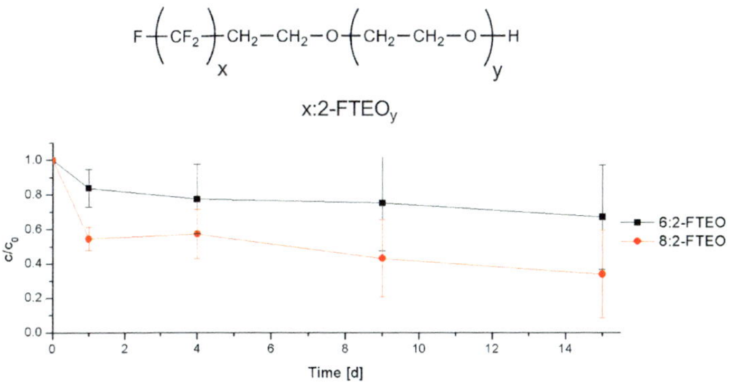

Fig. 1 Time course of concentration divided by the initial concentration of 6:2-FTEO (*black*) and 8:2-FTEO (*red*) congeners and their chemical structure. Please note that the *error bars* represent the standard deviation of ethoxymers within the homologue group [25]

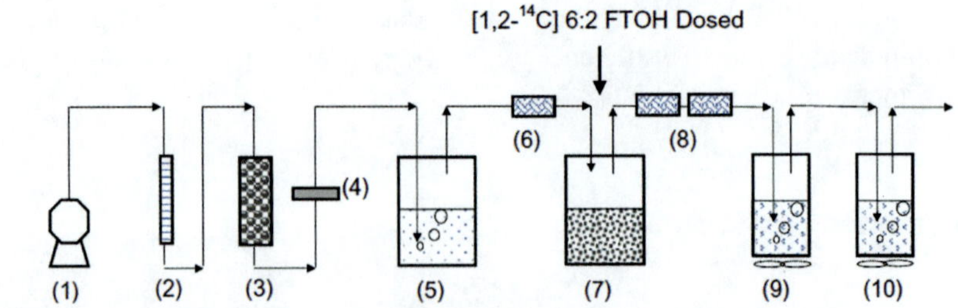

(1) Air pump (20 mL min^{-1}) (4) Air filter (0.45 μm) (7) Soil in 500-mL vessel

(2) Flow controller (5) Gas wash vessel (200 mL) (8) Two C_{18} cartridges at outlet

(3) Soda lime trap (40 g) (6) C_{18} cartridge at inlet (9) (10) NaOH solution traps (1 M)

Fig. 2 Sophisticated biodegradation experimental setup allowing for constant aeration and capturing volatile transformation products on C_{18} cartridges and basic traps. This setup was used to study biotransformation of $^{14}C_2$-labeled 6:2-FTOH. Taken from [29]

fluorotelomer ethoxylates (FTEO) is shown in a sterilized control experiment carried out in WWTP effluent using amber glass bottles. It can be seen that especially the 8:2-FTEO congeners are significantly adsorbed leading to approximately 50% of the initial concentration in solution after several days. In this case, it is not known, whether adsorption took place on the glassware or rather on particulate matter in the WWTP effluent.

In order to account for volatile products, closed-bottle conditions have been applied with few exceptions. Different techniques, such as solid-phase extraction (SPE) [16, 17, 21, 29, 31] or solid-phase microextraction [30] have been used to capture volatile metabolites. Recently, sophisticated biodegradation systems have been developed maintaining constant aerobic conditions and allowing to assess volatile metabolites [16, 29, 31] that can be easily stripped off the liquid phase (Fig. 2).

3.2 *Chromatographic Separations*

Chromatography describes a physico-chemical process, where a mixture of compounds is separated between a mobile and a stationary phase due to adsorption, partitioning or other effects. Although modern MS, especially tandem mass spectrometers (MS/MS), achieve an unprecedented selectivity, chromatography may become crucial when working with complex matrices. Such matrices often have to be dealt with when performing biotransformation studies, as these studies are often carried out in wastewater or soil. This requires proper separation of the compounds prior to MS.

While most analyses for PFC are performed using liquid chromatography (LC), gas chromatography (GC) still has a certain applicability for special purposes.

Well-known advantages of GC(-MS) over LC(-MS) are an unbeaten chromatographic resolution, which may be of importance for structural isomer separation, for example, for PFOS isomers [34], and less susceptibility to matrix effects. However, only a small fraction of PFC can be directly analyzed by GC methods, owing to the polar or even ionic structure of most of the PFC and their metabolites. Typical PFC that can be directly analyzed by GC are FTOH, fluorotelomer olefins, and other fluorotelomer-based compounds and metabolites [16]. However, the typical PFC such as PFCA and perfluorosulfonic acids (PFSA) are non-volatile and therefore not suited for GC analysis. This can be circumvented by derivatization, for example, to the butyl [34] or i-propyl esters [35] of sulfonates or preparation of the anilides [36] or methyl esters [37] from PFCA.

As stated above, most chromatographic separations of PFC are carried out by LC and exclusively under reversed-phase (RP) conditions [38]. Thus, PFC are retained basically by their perfluorocarbon chain length, but of course, functional groups attached to that moiety also influence the chromatographic behavior. Improved chromatographic selectivity compared with common C_{18} or C_8 phases can be achieved by special phases such as pentafluorophenyl (PFP) [39] or perfluorinated C_8 phases [16, 40], which both provide better selectivity for highly fluorinated substances and, in the latter case, circumvented false-positive results as compared with RP-C_{18} phases.

3.3 *Mass Spectrometry*

3.3.1 Atmospheric Pressure Ionization

Since the discovery of atmospheric pressure ionization (API) techniques was crucial for the investigations on PFC during the last decade, the functionality of Electrospray Ionization (ESI) – the most valuable API technique – will be presented in the following section.

The need to measure fluorinated and perfluorinated molecules arose already in the 1960s, when Taves discovered "two forms of fluorine in human serum" by ashing and subsequent potentiometric analysis with a fluoride-selective electrode [41]. During that time, however, no powerful tool that allows for sensitive and selective detection of these compounds was available. GC-MS, which was already available at that time, did not meet these criteria, mostly due to the ionic structure of the majority of PFC, which disallows volatilization needed for GC-MS.

This issue was solved only in the 1980s, when the group of John Fenn invented the ESI [42, 43] technique based on previous work by Dole and co-workers [44]. ESI allows for ionization and transfer into the gas phase of ionic compounds and macromolecules up to molar masses beyond 100 kDa. ESI-MS after LC has been routinely used ever since its commercialization [45].

ESI produces protonated, deprotonated or adduct ions (Na^+, K^+, NH_4^+, Cl^-, CH_3-COO^-, solvent clusters, etc.) by spraying a solution through an electrically charged capillary. The droplets formed contain a net charge and are accelerated toward a counter-electrode. Their size diminishes by evaporation, which may be thermally assisted. When a certain charge density is reached (the so-called Rayleigh limit), the droplets disintegrate by Coulombic repulsion leading to smaller droplets. This process is referred to as "Coulomb explosion." Formation of the free ions is explained by two different models: the "ion evaporation model" (IEM) by Iribarne and Thomson [46, 47] and the "charge residue model" (CRM) as proposed by Dole et al. [44]. Briefly, IEM suggests that ions are emitted from highly charged droplets into the gas phase, whereas with CRM ions are generated by complete evaporation of solvent resulting in free ions in the gas phase. While IEM holds for small inorganic ions, ionization of macromolecules such as proteins seems to be better explained by the CRM process. It is interesting that the ions observed in the gas phase, that is, those detected by the mass spectrometer, are not necessarily the same as those in solution [48]. This is only true for very stable ions, such as sodium ions. In the case of PFC, transfer from the liquid to the gas phase possibly only occurs for anions of very strong acids, such as PFSA, but possibly not for PFCA and related compounds, which may have higher pK_a values (although discussed very controversially, see [49–51]). This may seem odd, but is a direct consequence of the differences in liquid-phase and gas-phase acidity/basicity. To our knowledge, the mechanism of PFC ionization has not been investigated in detail. Some excellent reviews on the fundamentals of ESI have been published [48, 52–54].

Other API techniques, such as atmospheric pressure photoionization (APPI) and atmospheric pressure chemical ionization (APCI) have been marginally applied. Although providing advantages over ESI, such as reduced matrix effects, APCI has been rarely applied for PFC analysis. Analytes measured with APCI comprise various ethoxylated PFC [40, 55] and PFOA [56]. However, no investigations with respect to matrix effects were made in these articles.

A rather new technique is APPI, which has only been applied in two studies to determine PFOS [57] in river waters and FTOH and sulfonamido derivatives in biotic samples [58]. APPI is a very selective tool and, in stark contrast to ESI, is considered to be virtually imperceptible to matrix effects, which was confirmed in both studies. APCI and especially APPI are not recommended for metabolism studies of unknown compounds, since ionization is very delicate with these methods. Therefore, unknown compounds may not be discovered due to a lack of ionizability. ESI is the method of choice due to the wide range of ionizable compounds after LC separations.

3.3.2 Mass Analyzers

Today numerous mass analyzers are available with completely different characteristics and field of use. Both single-stage MS and multiple-stage MS instruments can be purchased commercially nowadays.

When performing GC-MS, a single quadrupole is still the mass analyzer routinely applied because of its very low cost in comparison with other MS instruments. A single quadrupole is rather unspecific, but due to very pronounced fragmentation of the compounds with electron impact (EI) ionization and the high chromatographic resolution of GC, selective analysis may be achieved.

Single quadrupole instruments can also be coupled to LC, but today, they have been largely replaced with triple quadrupole instruments (QqQ). They consist of two quadrupoles, which can be used for mass analysis separated by a collision cell, which is basically also a quadrupole (or hexa/octopole) that can be filled with an inert gas such as nitrogen or argon. By acceleration of the ions that pass the first quadrupole (so-called precursor ions), collision of these ions with the inert gas molecules or atoms can result in the formation of characteristic fragments, called product ions, which are then analyzed by the third quadrupole. QqQ instruments may be operated in different modes, which will be explained more thoroughly in the respective sections. These instruments are routinely used for trace analysis, since they are both very sensitive and selective due to multiple stage mass separation. More recent advances include exchange of Q3 by a linear ion trap (LIT). These instruments can also be used in "normal" quadrupole mode, thus offering the very same modes as QqQ instruments, but they may alternatively be run in advanced modes applying the LIT allowing for higher sensitivity, higher mass resolving power, and MS^3 scans.

Quadrupole ion traps (also referred to as "Paul Traps") can be used to generate product ion spectra. In fact, these instruments can offer MS^n scans with n up to 10 [59] allowing for investigation of fragmentation mechanisms and for thorough structural elucidation. Since separation of precursor and product ion(s) occurs at the same place but temporally shifted, ion trap MS/MS is referred to as "tandem mass spectrometry in time."

Other mass analyzers, especially those providing high-resolution, such as time-of-flight (ToF), Orbitrap and Fourier-transform ion cyclotron resonance MS, have been applied scarcely. Two biotransformation studies were carried out with a QqTOF-MS/MS, these advantages were used in order to gather accurate masses of metabolites [17, 60] and recently, the ultra-high resolution Orbitrap MS has been applied to confirm the presence of novel metabolites [21].

All abovementioned systems use collision-induced dissociation (CID), sometimes referred to as collision(ally)-activated dissociation (CAD), to form product ions of the species investigated. This process uses inert gas, mostly nitrogen or argon, to provoke collisions with accelerated ions that will lead to characteristic product ions. Unlike electron impact (EI) fragmentation, CID generally produces even-electron product ions, that is, ions with no unpaired electrons. This implies that neutral, non-radical species are generally cleaved off the precursor ions, often small organic or inorganic compounds, such as CO_2, HF, or H_2O. A great review on the detailed processes occurring during CID has been published by Levsen and Schwarz [61].

4 Approaches to Detect Novel Metabolites

4.1 General Modus Operandi

Detection of novel metabolites is a sophisticated task to solve, as the compounds are unknown and usually not main constituents of the mixture investigated. Since MS delivers both the mass of possible metabolites and valuable structural information, it is a perfect tool to accomplish such tasks.

Despite the high selectivity of MS, samples should always be compared with control samples, either not containing the test compound and/or with suppressed microbial activity. Otherwise, false-positive results are almost predestined. This is regardless of the instrumentation used for that purpose, since no mass spectrometer exhibits selectivity high enough to overcome potential interferences with environmental matrix compounds.

4.2 GC-MS

GC-MS is commonly delivered as single quadrupole instruments, although QqQ and ToF instruments are also commercially available.

With single quadrupole instruments, unknown components can only be detected in scan mode, where all *m*/*z* are recorded within a selected mass range. This mode offers good ability to detect unknowns because the background signal with GC-MS is usually very low, except for some siloxane fragments stemming from column bleeding. Thus, by comparing GC-MS chromatograms of samples and control samples, novel metabolites may be relatively easily detected.

However, most metabolites including those of fluorinated substances are ionic or at least highly polar, such as carboxylic acids. Thus, they are not compatible with GC analysis due to their non-volatility. Some fluorotelomer-based metabolites, such as fluorotelomer ketones [16] or secondary alcohols [17], are volatile and have been first described after detection by GC-MS.

4.3 LC-MS(/MS)

4.3.1 Single Quadrupole Instruments

A very straightforward way to detect unknowns is a single quadrupole scan recording all ions generated within a certain mass range. This is also possible with QqQ instruments by operating either Q1 or Q3 in scan mode and the other two in so-called "radio-frequency only" mode, which is equal to a bypass mode. Unlike for GC-MS, signal background with LC-MS techniques is usually very high (due to impurities in solvents and leaching of polymer additives from tubes) thus complicating the detection of non-target analytes.

There are several possibilities to resolve this problem by means of modern data-processing tools. The operator can extract certain ions equal to an assumed metabolite to produce "extracted ion chromatograms" (XIC), but then again, this is not a non-target analysis. More sophisticated data-processing tools are "base-peak chromatograms" (BPC) and isotopic pattern search tools. Whereas isotopic pattern search is not recommended with PFC due to the monoisotopic nature of fluorine, a BPC may be helpful. BPCs can be extracted by entering a certain mass range resulting in a new chromatogram, where the intensity of the most intense ion at a certain time is plotted against the time. This enhances visualization and facilitates detection of compounds in vast mixtures.

4.3.2 Triple Quadrupole Instruments

QqQ instruments offer two additional modes over single quadrupole instruments, which can be selectively used for detection of unknown compounds. Having in mind that these instruments consist of three quadrupoles (Q1, q2, and Q3) and that Q1 and Q3 can be used in both single ion monitoring (SIM) or scan mode, the functionalities can be altered according to the demand. Unfortunately, ways to detect unknown compounds are often neglected in the literature, where only positive findings and the confirmation of the presence of certain metabolites is presented. In this chapter, insight into potentially helpful procedures is given.

Modern instruments offer novel sophisticated modes to detect non-target analytes. If the fragmentation behavior of a compound class is known, this can be used to find derivatives of these compounds even though the structure of the complete molecule is unknown. If X represents the characteristic moiety of a compound class, which is attached to any moiety R, let us assume two hypothetic fragmentation pathways (fragmentation of anions is chosen because most PFC metabolites are ionized negatively):

$$R - X^{-} \rightarrow R + X^{-}, \tag{1}$$

$$R - X^{-} \rightarrow R^{-} + X. \tag{2}$$

In the first pathway, the characteristic moiety is cleaved off bearing the charge, whereas in the second pathway, the characteristic moiety X is cleaved off as a neutral compound. Clearly, these fragmentation pathways are facilitated. For instance, it occurs very frequently that moieties are cleaved off and attach or lose one hydrogen atom. Therefore, it is crucial to study the fragmentation of the specific compound class in advance.

Pathway (1) signifies that the moiety of interest forms a charged fragment. Thus, we would expect derivatives, like a biotransformation product, of an unknown molecular mass, to deliver the same product ion. Thus, Q1 is set to scan and Q3 is set to SIM, at the mass of the known product ion. This mode is widely known as "precursor ion scan."

Considering the second fragmentation pathway, the precursor ion scan is not applicable, since the product ion generated does not always have the same *m*/*z* ratio. However, the difference in *m*/*z* between the precursor ion and the production is equal. In order to detect such losses of a constant neutral fragment, both Q1 and Q3 are operated in scan mode, but at a constant *m*/*z* difference. Thus, a signal is only detected when the difference in *m*/*z* of the precursor ion and the product ion equals the desired value. This mode is commonly known as "neutral loss scan."

An example of the application of precursor ion scans is shown in Fig. 3. These data were recorded during a biodegradation study of a fluorosurfactant candidate containing the bis(trifluoromethyl)amino group. This group is cleaved off as the bis (trifluoromethyl)amide anion, $(CF_3)_2N^-$, at *m*/*z* 152. Thus, metabolites still carrying this group are also expected to form the same product ion. Considering the abovementioned explanations, this represents case (1), since the moiety of interest generates the charged fragment X^-.

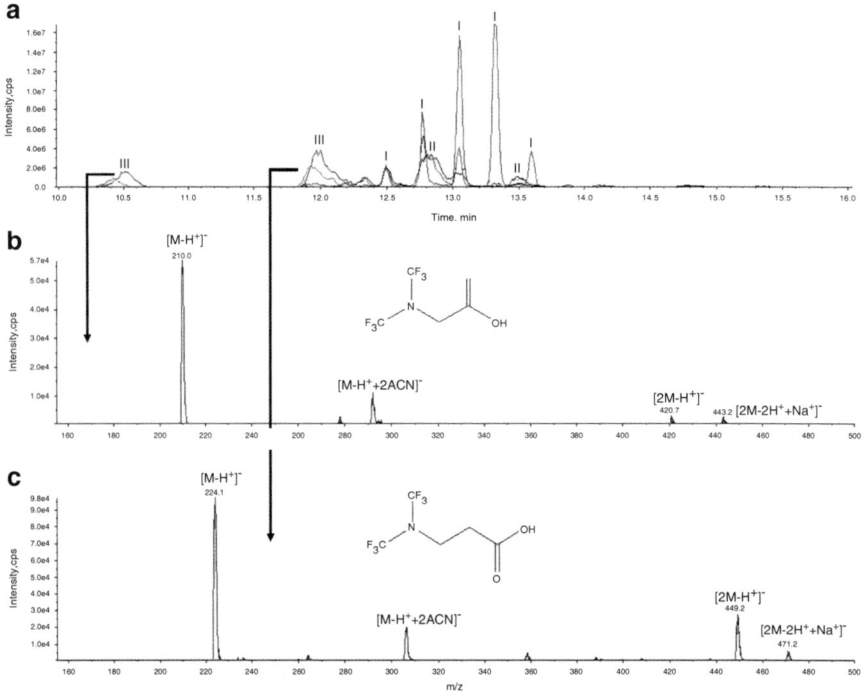

Fig. 3 LC-ESI-MS/MS precursor ion scans of biodegradation samples of ω-[Bis(trifluoromethyl) amino]-alkane-1-sulfonates. (**a**) TIC after 3 (*red*), 15 (*green*) and 34 days (*blue*). I = test compound, II = oxidized metabolites, III = carboxylic acid metabolites. (**b**) precursor ion scan spectrum at 10.5 min (**c**) precursor ion scan spectrum at 12 min. Adopted from [62]

The compound under investigation was a mixture of linear ω-[bis(trifluoromethyl)amino]alkane-1-sulfonates with a chain length distribution from 7 to 13 methylene groups. Samples analyzed were from three different dates, namely a sample after 3 days, where no transformation had occurred, and samples after 3 and 15 days, respectively [24, 62]. Only by recording precursor ion scans of these samples, a lot of information could be gathered in this case. First of all, the test substances are clearly visible as chromatographically separated peaks (red curve). It becomes evident that their peak areas diminish with increasing duration of the experiment, except for those eluting early ($t_R = 12.5$ min and approximately 12.75 min). These compounds bear a shorter alkyl chain than the others and were not degraded. Most importantly, a number of emerging peaks can be observed, which are not present in the chromatogram of day 0 ($t_R = 10.5$ min, 12 min, 12.9 min, 13.5 min). Recording a precursor ion scan, not only the total ion current (TIC) is measured over time (expressed here as the total ion chromatogram), but also mass spectra at Q1 are recorded, which allows to reveal the spectra behind the chromatographic peaks and thus determine the molecular mass of possible transformation products. By this means it was shown that two sets of metabolites were generated: oxidized metabolites, where one methylene group was oxidized to a carbonyl function (peaks labeled II at $t_R = 10.5$ min and 12 min), and a homologous series of ω-[bis(trifluoromethyl)amino]alkanoic acids (peaks labeled III). Of course, an identification of the presence of these compounds could not be achieved by these precursor ion scans, but their presence could be supposed. Identification by MS is discussed in the following section.

Another simple method to search for metabolites is to carry out SIM or MRM measurements of hypothetic metabolites. This is not a non-target analytical method as the previous ones, but offers greatly enhanced sensitivity and selectivity. However, care has to be taken when doing so: with single quadrupole SIM methods, the amount of "in-source fragmentation" has to be controlled by setting a "declustering potential" or "cone voltage" (depending on the MS manufacturer). Similarly, MRM methods are even more complicated to develop without proper authentic standards, as the number of voltages to be determined is even higher than in SIM mode and, even more delicately, product ions have to be known, which is often not trivial (see Sect. 5.3).

In all cases, potentially positive findings should be cross-checked with control samples. This is absolutely crucial to distinguish between matrix compounds and real metabolites. Given the fact that most biodegradation experiments are carried out in rather difficult matrices, such as sludge or soil, a high number of rather high-concentration matrix compounds is inherent. Especially for low-concentration biodegradation experiments, control samples are crucial. It is recommended to use at least sterile controls containing the test compound, inoculum, and a sterilizer (e.g., NaN_3, $HgCl_2$ or antibiotics) and non-sterile controls not containing the test compound in order to verify if degradation products of other compounds interfere with the analysis. Finally, the presence of assumed metabolites should be confirmed as presented in the following section.

5 Structure Determination of Unknown Compounds

5.1 General Modus Operandi

After a positive detection of a new metabolite, as presented in the previous section, the structure of this compound must be elucidated. Ideally, this should be carried out by comparison with an authentic standard, whose structure has been fully elucidated by MS and NMR methods, which was done by Wang et al. [16, 17].

However, this cannot always be performed due to a lack of such standards or difficulties in synthesizing them. Thus, in most cases, chemical structures can only be addressed by thorough investigation of the fragmentation patterns. These patterns will be described in this section after introduction of the MS modes used to gather MS spectra.

5.2 MS Modes Used to Determine Structures

Single quadrupole instruments are the simplest MS instruments and can only be used for structural elucidation with certain limitations. Due to only one stage of mass separation, chromatographic separation is the bottleneck of structure determination. If any compound coelutes with the target analyte, its signal will contribute to the mass spectrum and may thus lead to wrong assignments. Whereas GC separations usually yield very high chromatographic resolution and low MS background, this is not the case for LC separations. Therefore, LC/MS is not recommended to solve structural determination problems. In contrast, GC/MS has been applied several times to identify unknown metabolites of FTOH [16, 17, 30].

In order to overcome the abovementioned drawbacks, triple quadrupole MS should be applied after LC. This MS type is most commonly used to perform structural elucidation. They offer the well-known product ion scan mode (sometimes still referred to as daughter ion scan, albeit not recommended by the IUPAC [63]), where Q1 within a set of three quadrupoles is set to SIM of the respective precursor ion, that is, most likely the deprotonated metabolite, q2 is filled with inert gas to promote fragmentation by collisional activation, and Q3 scans the resulting product ions. Due to the spatially divided separation, this is referred to as tandem mass spectrometry in space. Product ion yield and pattern are highly depending on the collision energy (CE) applied. As a rule of thumb, higher CE leads to lower mass product ions. Novel improved instruments offer the functionality of a LIT instead of Q3, thus offering a product ion scan analog with a LIT collecting and analyzing the product ions (this is called "enhanced product ion scan," EPI). This allows for more sensitive detection and enhanced resolution. This mode has been used several times for structure determination of metabolites in our lab [24, 26]. Furthermore, it allows for the collection of MS scans fragmenting one selected product ion again, which can be very helpful for detailed structural analysis, for example, for structural isomers [24, 26].

In some very complicated cases, the structure of a metabolite may only be determined with the help of further methods, especially NMR. However, it has to be pointed out that such methods may require preparative chromatography or similar methods due to the relatively high amount of substance needed.

5.3 Fragmentation of Fluorinated Compounds

MS fragmentation of organic compounds is greatly influenced by the technique used. Fragmentation patterns occurring after EI are largely known and often follow distinct rules, such as α-cleavage or McLafferty rearrangement [64], which also hold for fragmentation of PFC. For CID, such general pathways do not exist and fragmentation differs largely from what has been known for EI fragmentation. Attempts have been made to study characteristic product ions or losses from organic compound classes [65], but these generalized pathways cannot be applied for every compound. This is a logical consequence of the different ion species produced: EI-MS initially leads to high-energy radical cations, whereas the ESI process – which is the most common ionization technique prior to CID – leads to protonated or deprotonated molecules. Thus, we need to focus on EI and CID fragmentation patterns separately.

When performing CID fragmentation with fluorinated compounds, unfortunately, there is no common fragmentation pathway that is significant for fluorine. Furthermore, very odd pathways may occur in the presence of fluorine, which will be subject of this section. Due to the unique properties of fluorine, fragmentation pathways of fluorinated molecules may differ largely from their non-fluorinated counterparts.

Fragmentation of classic PFCs, such as PFOS and PFOA has been studied very thoroughly [39, 66]. The PFSA are mainly known to produce sulfur-containing ions, such as SO_3^- (m/z 80) and FSO_3^- (m/z 99). However, this is only one part of the story. Technical mixtures of PFOS contain a number of positional isomers and the formation of the FSO_3^- ion is by far less abundant for these isomers as compared with non-branched PFOS [39]. Besides the two sulfur-containing ions, product ions of linear PFOS comprise – albeit to a low extent – perfluoroalkyl carbanions $C_mF_{2m+1}^-$ (the so-called "9-series," since the m/z values end with "9") as well as $C_nF_{2n}SO_3^-$ radical anions (the so-called "0-series") [39, 67]. The latter ones are suspected to be generated by initial radical cleavage of a C–C-bond within the perfluoroalkyl chain and subsequent losses of tetrafluoroethene. The perfluoroalkyl carbanions are supposed to derive from initial loss of SO_3 and subsequent loss of perfluoroalkenes such as tetrafluoroethene, hexafluoropropene, and so forth [67]. Interestingly, Langlois and Oehme found that the substitution site of trifluoromethyl-branched PFOS can be determined because of one missing "0-series"-ion in the spectrum, depending on the branching site.

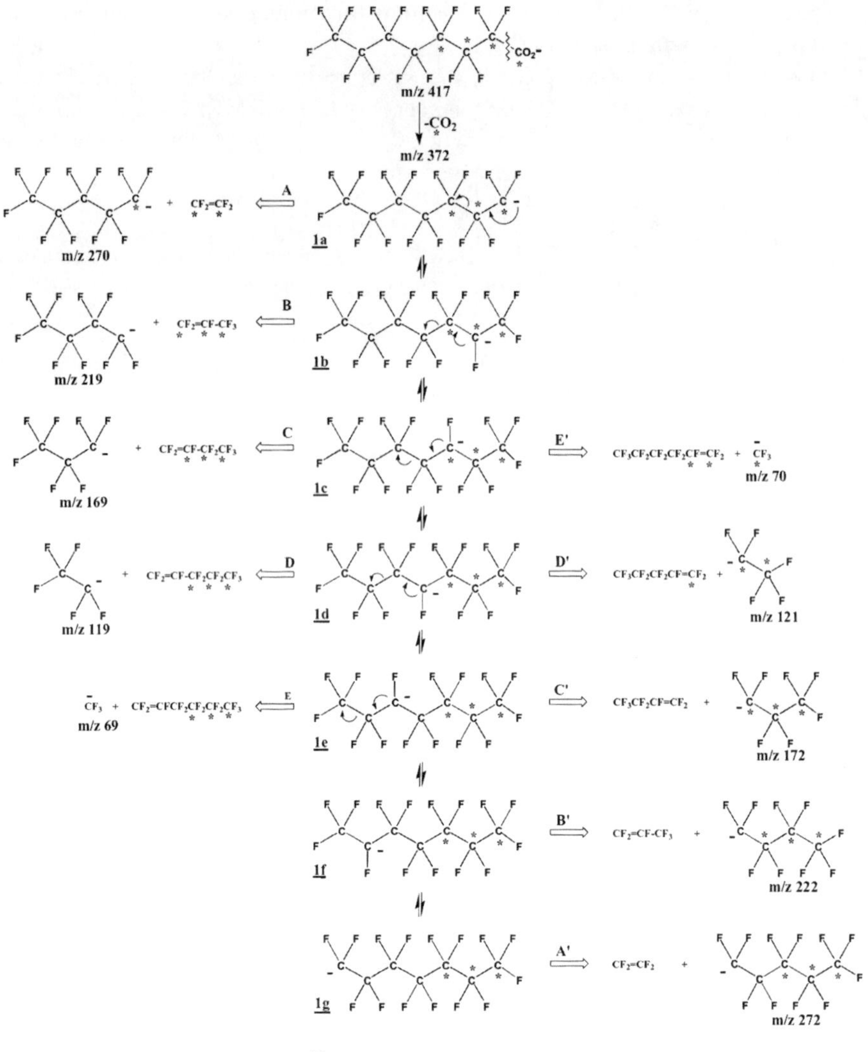

Fig. 4 Fragmentation pathway of $^{13}C_4$-PFOA including fluorine atom migration taken from [66]

Perfluorocarboxylate fragmentation is generally initiated by loss of CO_2, (Fig. 4) leading to a perfluoroalkyl carbanion $C_mF_{2m+1}^-$ which is normally only encountered with aromatic carboxylic acids, but not with aliphatic carboxylates [65]. This can be explained by stabilization of the negative charge by the proximity of the perfluoroalkyl chain or similar groups as in singly unsaturated perfluorinated alkenoic acids (Fig. 5).

Further fragmentation of perfluoroalkyl carbanions was studied very thoroughly by Arsenault et al. [66], who found that the initial formation of the respective $C_mF_{2m+1}^-$ ion is followed by fluorine atom migration and thus charge migration throughout the whole linear perfluoroalkyl chain (Fig. 4). They rationalize this

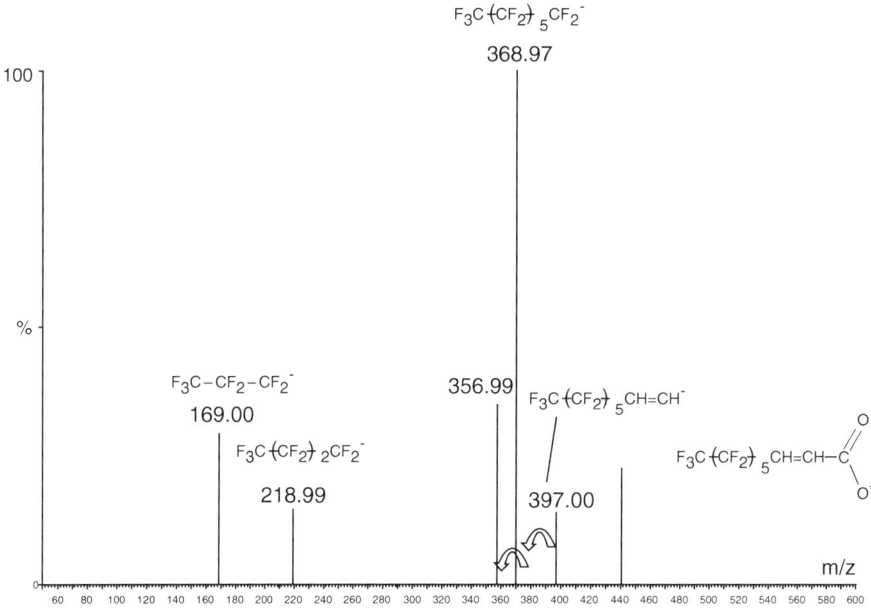

Fig. 5 Product ion spectrum of 7:3-fluorotelomer acid (7:3-FTUA). Arrows represent loss of HF. Adopted from [17], structures of product ions have been added by the authors

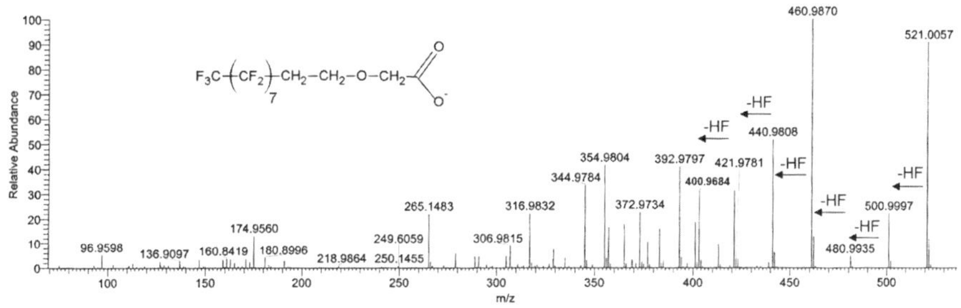

Fig. 6 Orbitrap spectrum of FTEO metabolite (so-called 8:2-FTEO1C) showing accurate masses of the fragments. Please note the striking fragmentation including six losses of HF, where all hydrogen atoms in the molecule are cleaved of as HF despite the apparent spatial distance between some hydrogen and fluorine atoms. This molecule also shows the fragment at *m/z* 355 characteristic of 8:2-fluorotelomer compounds

hypothesis by stating that the charge shift produces more stabilized secondary carbanions in contrast to the primary carbanions initially generated. Further fragmentation of these ions occurs by cleavage of different perfluoroalkenes. Similar fragmentation patterns are also observed for fluorotelomer acid derivatives, which also include loss of CO_2, and HF (Fig. 6).

Interestingly, if only one hydrogen atom is present in a perfluorinated alkyl chain, the fragmentation pathway may be altered substantially in contrast to the perfluorinated molecule. Under those circumstances, loss of hydrogen fluoride is often observed. A reason for this may be the energetically favored loss of HF in comparison to, for instance, loss of F_2, whose difference is energetically favored by 110 kcal/mol compared with F_2 [67]. Compounds falling under this category are 6:2-fluorotelomer sulfonates (6:2-FTS) [68] and 2H-PFOS [67]. Analogous to PFOS fragmentation, 6:2-FTS delivers SO_3^- (*m*/*z* 80) and HSO_3^- (*m*/*z* 81) as product ions.

Also FTOH may fragment by multiple loss of HF. However, FTOHs are very delicate species with respect to their ESI-MS performance. If only traces of organic anions, such as formiate or acetate are present, FTOHs will form adducts such as $[M + HCOO]^-$ or $[M + CH_3COO]^-$. Under exclusion of salts of these ions, the deprotonated molecule is formed [69]. It was discovered that MeOH favors its formation, whereas ACN inhibits it [70]. Additionally, addition of basic compounds such as ethanolamine can promote formation of the $[M–H]^-$ species [15]. Interestingly, it was discovered with the help of deuterated standards that the proton in vicinity to the perfluoroalkyl chain is cleaved off, not the hydroxyl proton, as one might expect. This again shows the strong negative inductive effect of perfluoroalkyl groups. Upon deprotonation, FTOHs suffer loss of all hydrogen atoms, although being separated by several bonds. A characteristic ion of fluorotelomer-derived compounds is the ion at *m*/*z* x55, where $x = 3$ for 8:2-fluorotelomer derivatives and $x = 4$ for 10:2-fluorotelomer derivatives and so forth. This ion was observed for fluorotelomer alcohols [69], fluorotelomer-based phosphates, so called PAPS and di-PAPS, for fluorotelomer ethoxylate (Rf-(CH_2-CH_2-O)$_n$-H and their metabolites (Rf-CH_2-CH_2-O-CH_2-COOH) (Frömel, unpublished work, see Fig. 6), but apparently not for FTS [68]. This suggests necessity and involvement of the Rf-CH_2-CH_2-O moiety.

In EI-MS, perfluoroalkyl chains often form fragment ions at *m*/*z* 69 $[CF_3]^+$, *m*/*z* 131 $[C_3F_5]^+$, *m*/*z* 169 $[C_3F_7]^+$ and so forth [16, 71]. Cleavage of a fluorine radical may also be observed leading to a loss of 19 Da [16]. EI of perfluorinated molecules at the common electron energy of 70 V generally only leads to the previously mentioned low-mass fragments and does not generate any observable molecular ion, which impairs usefulness of EI-MS for such purposes. Therefore, chemical ionization is often used to obtain information on the molecular weight of an unknown compound [16]. This shows that combined MS methods are best suited to determine the structure of new compounds.

References

1. Prevedouros K, Cousins IT, Buck RC et al (2006) Sources, fate and transport of perfluorocarboxylates. Environ Sci Technol 40:32–44
2. Paul AG, Jones KC, Sweetman AJ (2008) A first global production, emission, and environmental inventory for perfluorooctane sulfonate. Environ Sci Technol 43:386–392
3. Russell MH, Berti WR, Szostek B et al (2008) Investigation of the biodegradation potential of a fluoroacrylate polymer product in aerobic soils. Environ Sci Technol 42:800–807

4. Washington JW, Ellington JJ, Jenkins TM et al (2009) Degradability of an acrylate-linked, fluorotelomer polymer in soil. Environ Sci Technol 43:6617–6623
5. 3M (1999) Fluorochemical use, distribution and release overview; U.S. Public Docket AR-226-0550, http://www.chemicalindustryarchives.org/dirtysecrets/scotchgard/pdfs/226-0550.pdf, last access: 23 September 2009
6. Bondi A (1964) van der Waals volumes and radii. J Phys Chem 68:441–451
7. Robinson EA, Johnson SA, Tang TH et al (1997) Reinterpretation of the lengths of bonds to fluorine in terms of an almost ionic model. Inorg Chem 36:3022–3030
8. Gillespie RJ, Robinson EA (1992) Bond lengths in covalent fluorides. A new value for the covalent radius of fluorine. Inorg Chem 31:1960–1963
9. Brockway LO (1938) The internuclear distance in the fluorine molecule. J Am Chem Soc 60:1348–1349
10. Pauling L (1960) The nature of the chemical bond and the structure of molecules and crystals; an introduction to modern structural chemistry. Cornell University Press, Ithaca
11. Blanksby SJ, Ellison GB (2003) Bond dissociation energies of organic molecules. Acc Chem Res 36:255–263
12. Lemal DM (2004) Perspective on fluorocarbon chemistry. J Org Chem 69:1–11
13. Key BD, Howell RD, Criddle CS (1997) Fluorinated organics in the biosphere. Environ Sci Technol 31:2445–2454
14. Kissa E (2001) Fluorinated surfactants and repellents. Marcel Dekker, New York
15. Liu J, Lee LS (2007) Effect of fluorotelomer alcohol chain length on aqueous solubility and sorption by soils. Environ Sci Technol 41:5357–5362
16. Wang N, Szostek B, Buck RC et al (2009) 8-2 Fluorotelomer alcohol aerobic soil biodegradation: pathways, metabolites, and metabolite yields. Chemosphere 75:1089–1096
17. Wang N, Szostek B, Buck RC et al (2005) Fluorotelomer alcohol biodegradation – direct evidence that perfluorinated carbon chains breakdown. Environ Sci Technol 39:7516–7528
18. Arp HPH, Niederer C, Goss KU (2006) Predicting the partitioning behavior of various highly fluorinated compounds. Environ Sci Technol 40:7298–7304
19. Remde A, Debus R (1996) Biodegradability of fluorinated surfactants under aerobic and anaerobic conditions. Chemosphere 32:1563–1574
20. Rhoads KR, Janssen EML, Luthy RG et al (2008) Aerobic biotransformation and fate of N-ethyl perfluorooctane sulfonamidoethanol (N-EtFOSE) in activated sludge. Environ Sci Technol 42:2873–2878
21. Liu J, Wang N, Szostek B et al (2010) 6-2 Fluorotelomer alcohol aerobic biodegradation in soil and mixed bacterial culture. Chemosphere 78:437–444
22. Lange, CC (2000) The Aerobic Biodegradation of N-EtFOSE Alcohol by the Microbial Activity Present in Municipal Wastewater Treatment Sludge, http://www.nikwax.com/cmsdata/Downloads/pr/5-3M_biodegradation_report.pdf, last access: 23 September 2009
23. Boulanger B, Vargo JD, Schnoor JL et al (2005) Evaluation of perfluorooctane surfactants in a wastewater treatment system and in a commercial surface protection product. Environ Sci Technol 39:5524–5530
24. Frömel T, Peschka M, Fichtner N et al (2008) ω-(Bis(trifluoromethyl)amino)alkane-1-sulfonates: synthesis and mass spectrometric study of the biotransformation products. Rapid Commun Mass Spec 22:3957–3967
25. Frömel T, Knepper TP (2010) Fluorotelomer ethoxylates: Sources of highly fluorinated environmental contaminants – Part I: Biotransformation Chemosphere 80:1387–1392
26. Peschka M, Frömel T, Fichtner N et al (2008) Mechanistic studies in biodegradation of the new synthesized fluorosurfactant 9-[4-(trifluoromethyl)phenoxy]nonane-1-sulfonate. J Chromatogr A 1187:79–86
27. Peschka M, Fichtner N, Hierse W et al (2008) Synthesis and analytical follow-up of the mineralization of a new fluorosurfactant prototype. Chemosphere 72:1534–1540
28. Liu J, Lee LS, Nies LF et al (2007) Biotransformation of 8:2 fluorotelorner alcohol in soil and by soil bacteria isolates. Environ Sci Technol 41:8024–8030

29. Liu J, Wang N, Buck RC et al (2010) Aerobic biodegradation of [14C] 6:2 fluorotelomer alcohol in a flow-through soil incubation system. Chemosphere 80:716–723
30. Dinglasan MJA, Ye Y, Edwards EA et al (2004) Fluorotelomer alcohol biodegradation yields poly- and perfluorinated acids. Environ Sci Technol 38:2857–2864
31. Lee H, D'Eon J, Mabury SA (2010) Biodegradation of polyfluoroalkyl phosphates as a source of perfluorinated acids to the environment. Environ Sci Technol 44:3305–3310
32. Key BD, Howell RD, Criddle CS (1998) Defluorination of Organofluorine Sulfur Compounds by Pseudomonas Sp. Strain D2. Environ Sci Technol 32:2283–2287
33. Organisation of Economic Co-Operation and Development (2004) OECD Guideline for the Testing of Chemicals 309 – Aerobic Mineralisation in Surface Water
34. Chu S, Letcher RJ (2009) Linear and branched perfluorooctane sulfonate isomers in technical product and environmental samples by in-port derivatization-gas chromatography-mass spectrometry. Anal Chem 81:4256–4262
35. Langlois I, Berger U, Zencak Z et al (2007) Mass spectral studies of perfluorooctane sulfonate derivatives separated by high-resolution gas chromatography. Rapid Commun Mass Spec 21:3547–3553
36. Scott BF, Spencer C, Mabury SA et al (2006) Poly and perfluorinated carboxylates in North American precipitation. Environ Sci Technol 40:7167–7174
37. Moody CA, Field JA (1999) Determination of perfluorocarboxylates in groundwater impacted by fire-fighting activity. Environ Sci Technol 33:2800–2806
38. van Leeuwen SPJ, de Boer J (2007) Extraction and clean-up strategies for the analysis of poly- and perfluoroalkyl substances in environmental and human matrices. J Chromatogr A 1153:172–185
39. Langlois I, Oehme M (2006) Structural identification of isomers present in technical perfluorooctane sulfonate by tandem mass spectrometry. Rapid Commun Mass Spec 20:844–850
40. Schröder HF (2003) Determination of fluorinated surfactants and their metabolites in sewage sludge samples by liquid chromatography with mass spectrometry and tandem mass spectrometry after pressurised liquid extraction and separation on fluorine-modified reversed-phase sorbents. J Chromatogr A 1020:131–151
41. Taves DR (1968) Evidence that there are two forms of fluoride in human serum. Nature 217:1050–1051
42. Fenn JB, Mann M, Meng CK et al (1989) Electrospray ionization for mass spectrometry of large biomolecules. Science 246:64–71
43. Whitehouse CM, Dreyer RN, Yamashita M et al (1985) Electrospray interface for liquid chromatographs and mass spectrometers. Anal Chem 57:675–679
44. Dole M, Mack LL, Hines RL et al (1968) Molecular beams of macroions. J Chem Phys 49:2240–2249
45. Hansen KJ, Clemen LA, Ellefson ME et al (2001) Compound-specific, quantitative characterization of organic fluorochemicals in biological matrices. Environ Sci Technol 35:766–770
46. Iribarne JV, Thomson BA (1976) On the evaporation of small ions from charged droplets. J Chem Phys 64:2287–2294
47. Thomson BA, Iribarne JV (1979) Field induced ion evaporation from liquid surfaces at atmospheric pressure. J Chem Phys 71:4451–4463
48. Kebarle P, Peschke M (2000) On the mechanisms by which the charged droplets produced by electrospray lead to gas phase ions. Analytica Chimica Acta 406:11–35
49. Brace NO (1962) Long chain alkanoic and alkenoic acids with perfluoroalkyl terminal segments 1. J Org Chem 27:4491–4498
50. Burns DC, Ellis DA, Li H et al (2008) Experimental pKa determination for perfluorooctanoic acid (PFOA) and the potential impact of pKa concentration dependence on laboratory-measured partitioning phenomena and environmental modeling. Environ Sci Technol 42:9283–9288
51. Goss KU (2008) The pKa values of PFOA and other highly fluorinated carboxylic acids. Environ Sci Technol 42:456–458

52. Kebarle P, Tang L (1993) From ions in solution to ions in the gas phase – the mechanism of electrospray mass spectrometry. Anal Chem 65:972A–986A
53. Cech NB, Enke CG (2001) Practical implications of some recent studies in electrospray ionization fundamentals. Mass Spectrom Rev 20:362–387
54. Kebarle P, Verkerk UH (2009) Electrospray: from ions in solution to ions in the gas phase, what we know now. Mass Spectrom Rev 28:898–917
55. Schröder HF, Meesters RJ (2005) Stability of fluorinated surfactants in advanced oxidation processes – a follow up of degradation products using flow injection-mass spectrometry, liquid chromatography-mass spectrometry and liquid chromatography-multiple stage mass spectrometry. J Chromatogr A 1082:110–119
56. Sottani C, Minoia C (2002) Quantitative determination of perfluorooctanoic acid ammonium salt in human serum by high-performance liquid chromatography with atmospheric pressure chemical ionization tandem mass spectrometry. Rapid Commun Mass Spec 16:650–654
57. Takino M, Daishima S, Nakahara T (2003) Determination of perfluorooctane sulfonate in river water by liquid chromatography/atmospheric pressure photoionization mass spectrometry by automated on-line extraction using turbulent flow chromatography. Rapid Commun Mass Spec 17:383–390
58. Chu S, Letcher RJ (2008) Analysis of fluorotelomer alcohols and perfluorinated sulfonamides in biotic samples by liquid chromatography-atmospheric pressure photoionization mass spectrometry. J Chromatogr A 1215:92–99
59. Willoughby R, Sheehan E, Moitrovich S (2002) A global view of LC/MS. Global View Publishing, Pittsburgh, PA
60. Wang N, Szostek B, Folsom PW et al (2005) Aerobic biotransformation of 14C-labeled 8-2 telomer B alcohol by activated sludge from a domestic sewage treatment plant. Environ Sci Technol 39:531–538
61. Levsen K, Schwarz H (1983) Gas-phase chemistry of collisionally activated ions. Mass Spectrom Rev 2:77–148
62. Frömel T, Knepper TP (2008) Mass spectrometry as an indispensable tool for studies of biodegradation of surfactants. Trac Trend Anal Chem 27:1091–1106
63. Murray, KK, Boyd, RK, Eberlin, MN and others (2006) Standard Definitions of Terms relating to Mass Spectrometry (IUPAC Recommendations 2006), http://old.iupac.org/reports/provisional/abstract06/murray_prs.pdf, last access: 26 November 2010
64. McLafferty FW (1959) Mass spectrometric analysis. Molecular rearrangements. Anal Chem 31:82–87
65. Levsen K, Schiebel H-M, Terlouw JK et al (2007) Even-electron ions: a systematic study of the neutral species lost in the dissociation of quasi-molecular ions. J Mass Spectrom 42:1024–1044
66. Arsenault G, McAlees A, McCrindle R et al (2007) Analysis of perfluoroalkyl anion fragmentation pathways for perfluoroalkyl carboxylates and sulfonates during liquid chromatography/tandem mass spectrometry: evidence for fluorine migration prior to secondary and tertiary fragmentation. Rapid Commun Mass Spec 21:3803–3814
67. Lyon PA, Tomer KB, Gross ML (1985) Fast atom bombardment and tandem mass spectrometry for characterizing fluoroalkanesulfonates. Anal Chem 57:2984–2989
68. Schultz MM, Barofsky DF, Field JA (2004) Quantitative determination of fluorotelomer sulfonates in groundwater by LC MS/MS. Environ Sci Technol 38:1828–1835
69. Berger U, Langlois I, Oehme M et al (2004) Comparison of three types of mass spectrometer for high-performance liquid chromatography/mass spectrometry analysis of perfluoroalkylated substances and fluorotelomer alcohols. Eur J Mass Spectrom 10:579–588
70. Szostek B, Prickett KB, Buck RC (2006) Determination of fluorotelomer alcohols by liquid chromatography/tandem mass spectrometry in water. Rapid Commun Mass Spec 20:2837–2844
71. Huang S (1989) Obtaining structural information on perfluorocarbons by mass spectrometry II – charge-exchange method. Org Mass Spectrom 24:1065–1071

Sorption and Leaching Behavior of Perfluorinated Compounds in Soil

Vanessa Gellrich and Thomas P. Knepper

Abstract Perfluorinated compounds can be detected worldwide in both, soil and water. In order to study the sorption and leaching behavior of this heterogeneous group of compounds in soil, among others, flow-through column experiments have been conducted. These experiments performed so far show that the percolation velocity is strongly dependent on the size i.e., the chain length of the molecule. Perfluorinated compounds with short chain lengths leach faster than perfluorinated compounds with longer chain lengths. Other factors that may influence the leaching behavior are the functional group of the perfluorinated compounds, the organic carbon content of the soil and the presence of other adsorbates. The dominating perfluorinated compounds in surface waters are perfluorooctanoic acid and perfluorooctane sulfonic acid. With these data it will be possible to model the environmental fate of perfluorinated compounds of different chain lengths.

Keywords Groundwater • Leaching • Perfluorobutanoic acid • PFC • Sorption

Contents

V. Gellrich • T.P. Knepper (✉)
University of Applied Sciences Fresenius, Institute for Analytical Research, Limburger Str. 2, 65510 Idstein, Germany
e-mail: knepper@hs-fresenius.de

T.P. Knepper and F.T. Lange (eds.), *Polyfluorinated Chemicals and Transformation Products*, Hdb Env Chem (2012) 17: 63–72,
DOI 10.1007/978-3-642-21872-9_4,

Abbreviations

Kd	Partition or distribution coefficient
K_{OC}	Soil organic carbon normalized distribution coefficient
PFBA	Perfluorobutanoic acid
PFBS	Perfluorobutane sulfonic acid
PFC	Per- and polyfluorinated compounds
PFCA	Perfluorocarboxylic acid
PFDA	Perfluorodecanoic acid
PFDoDA	Perfluorododecanoic acid
PFDS	Perfluorodecane sulfonic acid
PFHpA	Perfluoroheptanoic acid
PFHxA	Perfluorohexanoic acid
PFHxS	Perfluorohexane sulfonic acid
PFNA	Perfluorononanoic acid
PFOA	Perfluorooctanoic acid
PFOS	Perfluorooctane sulfonic acid
PFOSA	Perfluorooctane sulfonamide
PFPeA	Perfluoropentanoic acid
PFSA	Perfluoro sulfonic acid
PFTeDA	Perfluorotetradecanoic acid
PFUnDA	Perfluoroundecanoic acid
WWTP	Waste water treatment plant

1 Introduction

The behavior and transport of Per- and polyfluorinated compounds (PFC) in the environment has been discussed variously with the aim of understanding how these substances of industrial origin can reach remote areas. In 2001 Giesy found PFOS in high concentrations in arctic mammals and polar bears [1], suggesting that the marine food chain was contaminated via distribution of PFC in ocean currents. In 2004, Ellis et al. [2] proposed volatile fluorotelomer alcohols (FTOH) as precursors to a homologous series of PFCA. Model-based evaluations of the major transport ways indicated that the oceanic transport of PFCA is much more important than the atmospheric degradation of FTOH in delivering PFCA to the Arctic [3, 4].

Figure 1 shows how PFC can spread in the environment. When the discharge of PFC into the environment occurs at point sources (e.g., after the use of fire fighting foams), large amounts of PFC may be released. However, diffuse sources also release considerable amounts of PFC into the environment [5, 6]. In waste water treatment plants (WWTP), PFC can accumulate in sewage sludge [7–9]. As this sludge is used as fertilizer or soil conditioner, the accumulated PFC are applied to fields. The concentrations of PFOS and PFOA in sewage sludge are already regulated in some countries before it may be used as fertilizer. However, large

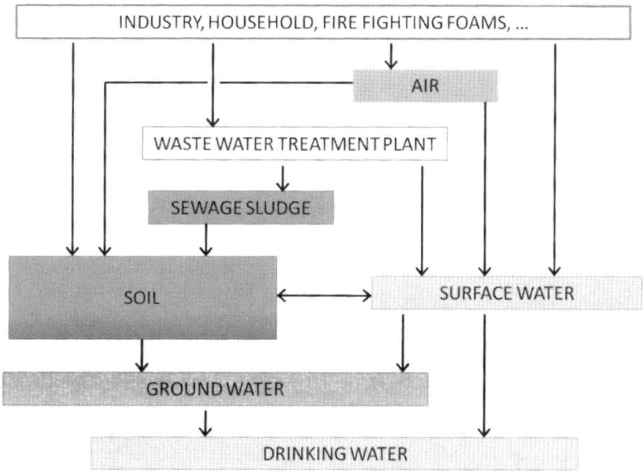

Fig. 1 Emission pathways of PFC in the environment

concentrations of other PFC or precursors, which can degrade on the fields or in the WWTP, can be found in the sludge [10, 11]. It has been shown that PFC can leach in considerable amounts from soil to ground and surface waters [12–14], what might be of concern for drinking water abstraction.

The sorption- and remobilization potentials of the different PFC may be essential for the assessment of the acute and long-term exposure of ground and drinking water to PFC. PFC are readily water soluble [15], but they also adsorb onto, or accumulate in, solid matrices such as soil, sediments, plants and animals [16]. As PFC can migrate from soil to plants [17, 18] the behavior of PFC in soil has also impact on their occurrence in field crop and the food chain. Food and especially drinking water are considered as the major sources of PFC found in the human body [19–22].

2 Sorption and Soil Passage

When examining the properties of PFC, field data as well as results of laboratory experiments can give important information.

2.1 Laboratory Sorption Experiments

Higgins and Luthy [23] examined the organic carbon normalized distribution coefficients (K_{OC}) of eight long chain PFC (four PFCA, two PFSA and two perfluorooctane sulfonamido acetic acids) with batch experiments (initial PFC concentrations 0.5–100 μg/L) (results see Table 1). They displayed the dependency of sorption and chain length, showing that the log K_{OC} values increase by between 0.3 and 0.6 log units with each additional CF_2 moiety. When comparing the results

Table 1 Summary of distribution coefficients

–	PFOA	PFNA	PFDA	PFUnDA	PFBS	PFOS	PFDS	Ref.
	2.06	2.39	2.76	3.30	n.d.	2.57	3.53	Higgins and Luthy 2006 [23]
	0.04–0.63	0.62–1.26	1.45–1.90	n.d.	−0.39–0.70	1.18–1.60	n.d.	Enevolsen et al. 2010 [24]
	n.d.	n.d.	n.d.	n.d.	n.d.	2.4–2.6	n.d.	Johnson et al. 2007 [25]
log K_{OC}	2.63	3.69	n.d.	n.d.	n.d.	3.16	n.d.	Kwadijk et al. 2010 [34]
	1.83	2.89	2.87	n.d.	1.42	2.35	n.d.	Kwadijk et al. 2010 [34]
	0.04–0.26	0.62–0.89	1.45–1.52	n.d.	−0.39–1.15	1.18–1.23	n.d.	Enevolsen et al. 2010 [24]
	n.d.	n.d.	n.d.	n.d.	n.d.	0.45–0.95	n.d.	Johnson et al. 2007 [25]
	−0.22–0.30	n.d.	n.d.	n.d.	n.d.	0.48–0.97	n.d.	Barkowski et al. [13]
log Kd	n.d.	n.d.	n.d.	n.d.	n.d.	0.87–1.55	n.d.	3 M 2003 [26]

n.d.: not determined

of sulfonic acids and carboxylic acids with the same number of perfluorinated carbon atoms, the sulfonic acids adsorb more (about 0.2 log units). They also found that the sediment sorption kinetics is slow. It took 10 days to reach equilibrium. Two types of interactions with the sediment seem to be important. Hydrophobic interactions between the organic carbon content of the sediment and the perfluorinated carbon chain on the one hand and electrostatic interactions affected by the functional group on the other hand. With decreasing pH and increasing Ca^{2+} concentration an increase of sorption could be measured.

Comparable batch experiments were conducted by Enevolsen et al. [24], but they used top soils instead of freshwater sediments and chose a higher Ca^{2+} concentration (100 mM $CaCl_2$ instead of 0.5 mM). They determined Kd and K_{OC} values and also did desorption experiments. Their initial concentrations were between 0.02 and 1 μg/L. Their distribution coefficients were lower than those obtained by Higgins et al. (Table 1) but they also found a correlation between the log K_{OC} and the molecular weight, and the sulfonic acids showed higher sorption than their corresponding carboxylic acids, too. The desorption was lower than adsorption, indicating that soil might act as a protective barrier towards groundwater contamination.

Johnson et al. [25] tested the sorption behavior by equilibrating five materials with solutions of PFOS (high initial concentrations were between 0.12 and 8 mg/L). The Kd values ranged from 2.8 to 8.9 L/kg (see Table 1). They also found a decrease in adsorption with increasing pH and also suggested that both, the inorganic and the organic content of the sediment play an important role in the sorption process.

3 M studied the properties of PFOS in 2003 [26] and found a rapid adsorption (equilibrium in less than 24 h) and Kd values between 7.4 and 35 L/kg for different

soils and sediments. The authors assumed that, despite this strong adsorption, PFOS would be mobile in the aqueous phase at equilibrium.

Tang et al. [27] performed batch sorption experiments with PFOS on goethite (FeO(OH)) and silica as model adsorbents of geo-environmental significance (PFOS initial concentration was 5–1,000 μg/L). They found only a marginal effect of pH, ionic strength or calcium concentration for the sorption to the negatively charged silica, indicating non-electrostatic interactions, but strong electrostatic interaction with goethite due to the positively charged surface.

2.2 Sorption and Distribution in the Environment

In a review of the maximum reported PFC concentrations made by Rayne and Forest [28], ground water was the only aquatic matrix where no PFC with a carbon chain longer than ten was reported. In the other matrices, such as lake, river, drinking or waste water, PFC up to 12 C-atoms could be measured in low concentrations. For the PFCA, PFOA was, in most matrices, the substance with the highest concentrations, whereas for the PFSA, the concentrations of PFBS and PFHxS often were as high as for PFOS. PFC with longer carbon chains can rather be found in solid matrices [16, 29–33].

Ahrens et al. [34] investigated the distribution of 40 different short and long chain PFC in the dissolved phase and in suspended particulate matter in the River Elbe. The total riverine PFC flux was estimated to be 802 kg/year for the dissolved phase and 152 kg/year for the particulate phase. Most PFC could not be identified in the particulate phase but only in the dissolved phase. PFOA showed the highest concentration in the dissolved phase (up to 12.5 ng/L) and PFOSA and PFOS were the predominant PFC in the particulate matter. In another study [33], they analyzed sediment cores and the pore water and found a very strong influence of the perfluorocarbon chain length and functional group on the partitioning behavior. Short chain PFCA ($C < 7$) could be found exclusively in pore water and long chain PFCA ($C > 11$) were found only in sediment. In general, PFCA could be normally found in pore water whereas PFSA (PFHxS and PFOS) were predominantly adsorbed to the sediment. They also confirmed the findings of Higgins and Luthy [23] that the sorption of PFC increases with increasing organic matter and decreasing pH.

Becker et al. [35] measured the concentration of PFOA and PFOS in river water and sediment upstream and downstream a WWTP. In water, the concentrations for PFOA (10–23 ng/L) were higher than for PFOS (2–16 ng/L) whereas in sediment, the PFOS concentrations were much higher (72–300 ng/kg compared with 18–68 ng/kg for PFOA). This again indicates a much higher sorption potential of PFOS.

Kwadijk et al. [36] calculated distribution coefficients (Kd and K_{OC}) and bioaccumulation factors for five PFC from field data gained from monitoring eel, sediment and water from the Netherlands (see Table 1). PFC concentrations were in the low ng/g (max. 13 ng/g for PFBS) or ng/L range (max. 43 ng/L for PFOA, max. 32 ng/L for PFOS and max. 290 ng/L for PFBS). Unlike the findings of Higgins and Luthy, their calculated Kd and K_{OC} values were higher for carboxylic acids than for sulfonates (log $K_{OC} = 3.69 \pm 0.52$ for PFNA vs. log $K_{OC} = 3.16 \pm 0.28$ for PFOS).

2.3 Soil Passage

Moody and Field [12] identified PFCA in groundwater under fire-training areas, even 10 years after the last use of fire fighting foams. This observation confirmed the possibility that PFC can reach the groundwater in considerable concentration (up to 6.6 mg/L PFOA) and that PFC are relatively immune to biodegradation.

Initial investigations of Barkowski et al. [13] concerning the sources and impacts of PFC in North Rhine-Westphalia revealed a precipitation depending mobilization of PFC. Soil and water affected by the application of a highly contaminated soil conditioner were examined. With increasing rainfall, increasing transport and concentration of PFOA in the rivers was measured. Additionally they carried out leaching experiments with two authentic soils from the area Brilon-Scharfenberg (PFOA concentration 0.4–0.7 mg/kg and PFOS concentration 1.5–6.6 mg/kg) and calculated Kd values from these results (see Table 1). The Kd values indicate a much higher mobilization potential for PFOA than for PFOS. The leaching of PFOS was significantly slower than the leaching of PFOA. This agrees with monitoring data, which showed that PFOA concentrations in ground water, surface and drain water were more than five times higher than the PFOS concentrations although the soil PFOS concentrations were about seven times higher than for PFOA.

With the aim of reproducing natural conditions, Murakami et al. [5] performed soil infiltration column tests with artificial street runoff, which was fed continuously through a loamy soil (PFC initial concentrations ranged between about 1 ng/L for PFUnDA and 47 ng/L for PFOA). Depending on the test conditions, up to 20% of PFOA and PFNA were removed by the soil, more than 60% of PFOS and PFDA and more than 80% of PFOSA and PFUnDA. Data from a monitoring study of groundwater, rivers, wastewater and street runoff supported the view that the efficiency of removal during infiltration increased with the chain length.

Leaching experiments were performed by Gellrich et al. [14] to gather information about the sorption and desorption behavior of PFC of different chain lengths in soil. Water-saturated loamy sand was spiked with aqueous PFC solutions or with contaminated sewage sludge, leading to concentrations of 10 μg/kg soil to 1 mg/kg soil. By analyzing the percolating water of this flow-through column experiment, the mobility of the different PFC could be assessed. Figure 2 shows breakthrough curves of columns spiked with a mixture of 14 PFC. A dependency of the leaching behavior to the chain length can be seen. The short chain PFC elute without retention but PFC with eight or more fluorinated carbon atoms could not be detected in the percolate after 2 years.

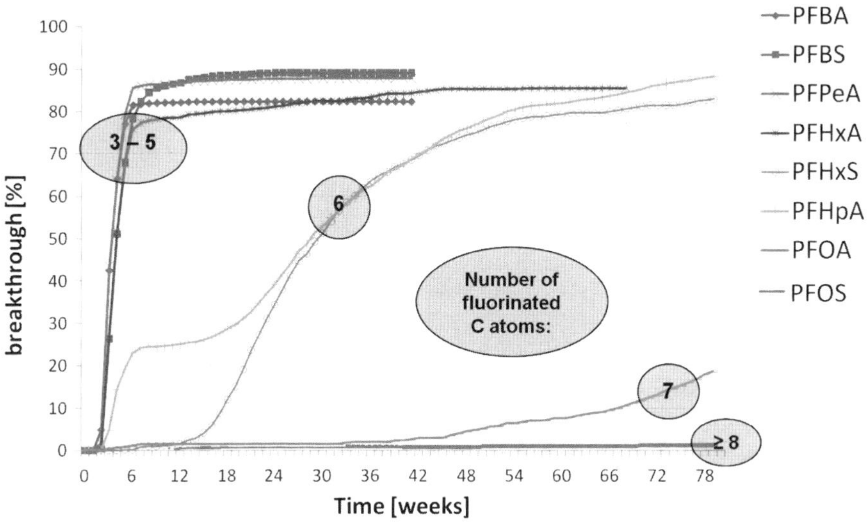

Fig. 2 Breakthrough curves of columns spiked with a mixture of 14 PFC, each 2 μg; $n = 2$

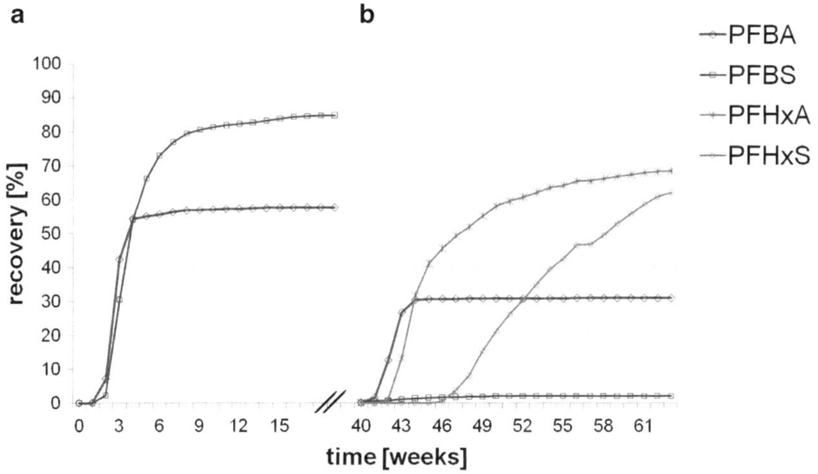

Fig. 3 (**a**): *Breakthrough curves* of columns spiked with PFBA and PFBS; (**b**): week 40: addition of PFHxA and PFHxS; 10 μg of each PFC; $n = 2$

When only PFBA and PFBS were spiked, PFBA did not elute completely. About 40% of the added PFBA appeared to be irretrievable (Fig. 3a). After adding PFC with a longer carbon chain, the "lost" amount of PFBA began to elute again (Fig. 3b). The same phenomenon was observed after adding stearate. Thus it would appear that larger and more lipophilic molecules (here PFHxA or PFHxS and stearate) can displace shorter PFC (here PFBA) from their binding sites in the soil. Other factors that may influence the leaching behavior are the functional group of the PFC, the organic carbon content of the soil and the flow rate of the percolating water.

3 Summary and Conclusion

Different studies have shown that PFC undergo at least two different types of interactions with the adsorbent. A hydrophobic interaction of the perfluorinated carbon tail (e.g., with the organic carbon fraction of the soil) and an electrostatic interaction of the head group (e.g., to the charged clay fraction of the soil). To express the sorption behavior, distribution coefficients such as the Kd or the K_{OC} are used. The use of K_{OC} is probably not the best descriptor for thc sorption behavior of PFC as it does not take all possible influences into account.

The dependency of the physico-chemical properties on the chain length causes a different distribution pattern for the different PFC. The longer the perfluorinated carbon chain, the higher the bio-concentration factors and the higher the tendency to adsorb to solid matrices. Thus short chain PFC are more likely to be found in aqueous matrices whereas long chain PFC are predominantly in solid matrices.

The specific sorption properties can be used to remove PFC from water, but regarding groundwater contamination and the accumulation of PFC in our food chain, the behavior and distribution of PFC in the environment is an important issue we still have to learn more about.

References

1. Giesy JP, Kannan K (2001) Global distribution of perfluorooctane sulfonate in wildlife. Environ Sci Technol 35:1339–1342
2. Ellis DA, Martin JW, de Silva AO et al (2004) Degradation of fluorotelomer alcohols: a likely atmospheric source of perfluorinated carboxylic acids. Environ Sci Technol 38:3316–3321
3. Wania F (2007) A global mass balance analysis of the source of perfluorocarboxylic acids in the Arctic Ocean. Environ Sci Technol 41:4529–4535
4. Armitage JM, MacLeod M, Cousins IT (2009) Comparative assessment of the global fate and transport pathways of long-chain perfluorocarboxylic Acids (PFCAs) and perfluorocarboxylates (PFCs) emitted from direct sources. Environ Sci Technol 43:5830–5836
5. Murakami M, Kuroda K, Sato N et al (2009) Groundwater pollution by perfluorinated surfactants in Tokyo. Environ Sci Technol 43:3480–3486
6. Prevedouros K, Cousins IT, Buck RC et al (2006) Sources, fate and transport of perfluorocarboxylates. Environ Sci Technol 40:32–44
7. Becker AM, Gerstmann S, Frank H (2008) Perfluorooctane surfactants in waste waters, the major source of river pollution. Chemosphere 72:115–121
8. Bossi R, Strand J, Sortkjaer O et al (2008) Perfluoroalkyl compounds in Danish wastewater treatment plants and aquatic environments. Environ Int 34:443–450
9. Higgins CP, Field JA, Criddle CS et al (2005) Quantitative determination of perfluorochemicals in sediments and domestic sludge. Environ Sci Technol 39:3946–3956
10. Sinclair E, Kannan K (2006) Mass loading and fate of perfluoroalkyl surfactants in wastewater treatment plants. Environ Sci Technol 40:1408–1414

11. Rhoads KR, Janssen EML, Luthy RG et al (2008) Aerobic biotransformation and fate of N-ethyl perfluorooctane sulfonamidoethanol (N-EtFOSE) in activated sludge. Environ Sci Technol 42:2873–2878
12. Moody CA, Field JA (1999) Determination of perfluorocarboxylates in groundwater impacted by fire-fighting activity. Environ Sci Technol 33:2800–2806
13. Barkowski D (2007) Pilotuntersuchungen zu Vorkommen und Auswirkungen von perfluorierten Tensiden (PFT) in Abfällen, die der BioAbfV unterliegen. Abschlussbericht. Unter Mitarbeit von Petra Günther, Florian Raecke und Detlef Wind. Herausgegeben von Institut für Umwelt-Analyse Projekt GmbH. Bezirksregierung Arnsberg
14. Gellrich V, Stahl T, Knepper T (2011) Investigating the fate of perfluorinated compounds in soils with leaching experiments (in press)
15. Kissa E (2001) Fluorinated surfactants and repellents. 2. ed., rev. and expanded. Marcel Dekker, New York, Surfactant Science Series, 97
16. Dietz R, Bossi R, Riget FF et al (2008) Increasing perfluoroalkyl contaminants in east Greenland polar bears (Ursus maritimus): A new toxic threat to the Arctic bears. Environ Sci Technol 42:2701–2707
17. Stahl T, Heyn J, Thiele H et al (2009) Carryover of perfluorooctanoic acid (PFOA) and perfluorooctane sulfonate (PFOS) from soil to plants. Arch Environ Contam Toxicol 57: 289–298
18. Weinfurter K, Kördel W, Bücking M (2008) Untersuchungen zum Übergang von PFT aus belasteten Böden in Pflanzen. Bodenschutz: 88–92
19. Ericson I, Nadal M, van Bavel B et al (2008) Levels of perfluorochemicals in water samples from Catalonia.Spain: is drinking water a significant contribution to human exposure. Environ Sci Pollut Res 15:614–619
20. Ericson Jogsten I, Perello G, Llebaria X et al (2009) Exposure to perfluorinated compounds in Catalonia, Spain, through consumption of various raw and cooked foodstuffs, including packaged food. Food Chem Toxicol 47:1577–1583
21. Fromme H, Tittlemier SA, Völkel W et al (2009) Perfluorinated compounds – exposure assessment for the general population in western countries. Int J Hyg Environ Health 212: 239–270
22. Trudel D, Horowitz L, Wormuth M et al (2008) Estimating consumer exposure to PFOS and PFOA. Risk Anal 28:251–269
23. Higgins CP, Luthy RG (2006) Sorption of perfluorinated surfactants on sediments. Environ Sci Technol 40:7251–7256
24. Enevoldsen R, Juhler RK (2010) Perfluorinated compounds (PFCs) in groundwater and aqueous soil extracts: using inline SPE-LC-MS/MS for screening and sorption characterisation of perfluorooctane sulphonate and related compounds. Anal Bioanal Chem 398:1161–1172
25. Johnson RL, Anschutz AJ, Smolen JM et al (2007) The adsorption of perfluorooctane sulfonate onto sand, clay, and iron oxide surfaces. J Chem Eng Data 52:1165–1170
26. 3M Company (2003) Environmental and health assessment document for perfluorooctane sulfonate and its salts
27. Tang CY, Fu QS, Gao D et al (2010) Effect of solution chemistry on the adsorption of perfluorooctane sulfonate onto mineral surfaces. Water Res 44:2654–2662
28. Rayne S, Forest K (2009) Perfluoroalkyl sulfonic and carboxylic acids: A critical review of physicochemical properties, levels and patterns in waters and wastewaters, and treatment methods. J Environ Sci Health A Tox Hazard Subst Environ Eng 44:1145–1199
29. Conder JM, Hoke RA, de Wolf W et al (2008) Are PFCAs bioaccumulative? A critical review and comparison with regulatory lipophilic compounds. Environ Sci Technol 42:995–1003
30. Butt CM, Mabury SA, Kwan M et al (2008) Spatial trends of perfluoroalkyl compounds in ringed seals (Phoca hispida) from the Canadian Arctic. Environ Toxicol Chem 27:542–553
31. Alzaga R, Salgado-Petinal C, Jover E et al (2005) Development of a procedure for the determination of perfluorocarboxylic acids in sediments by pressurised fluid extraction,

headspace solid-phase microextraction followed by gas chromatographic-mass spectrometric determination. J Chrom A 1083:1–6
32. Washington JW, Henderson WM, Ellington JJ et al (2008) Analysis of perfluorinated carboxylic acids in soils II: Optimization of chromatography and extraction. J Chrom A 1181:21–32
33. Ahrens L, Yamashita N, Yeung LWY et al (2009) Partitioning behavior of per- and polyfluoroalkyl compounds between pore water and sediment in two sediment cores from Tokyo Bay, Japan. Environ Sci Technol 43:6969–6975
34. Ahrens L, Plassmann M, Xie Z et al (2009) Determination of polyfluoroalkyl compounds in water and suspended particulate matter in the river Elbe and North Sea Germany. Front Environ Sci Eng China 3:152–170
35. Becker AM, Gerstmann S, Frank H (2008) Perfluorooctanoic acid and perfluorooctane sulfonate in the sediment of the Roter Main River, Bayreuth, Germany. Environ Pollut 156:818–820
36. Kwadijk CJAF, Korytàr P, Koelmans AA (2010) Distribution of perfluorinated compounds in aquatic systems in The Netherlands. Environ Sci Technol 44:3746–3751

Polyfluorinated Chemicals in European Surface Waters, Ground- and Drinking Waters

Christian Eschauzier, Pim de Voogt, Heinz-Jürgen Brauch, and Frank Thomas Lange

Abstract Polyfluorinated chemicals (PFCs), especially short chain fluorinated alkyl sulfonates and carboxylates, are ubiquitously found in the environment. This chapter aims at giving an overview of PFC concentrations found in European surface, ground- and drinking waters and their behavior during conventional water treatment steps.

Main PFC sources to the aquatic environment are municipal and industrial wastewater treatment plants. Treated landfill leachates also showed to be an important source of PFCs to surface waters. Existing data suggest central and south European rivers to have higher concentrations and mass discharges of PFCs than Northern European countries. However, this conclusion might be an artifact due to differences of monitoring activities in different regions.

High PFC levels in groundwater are often restricted to some contaminated areas, e.g., due to illegal waste deposition on agricultural land or in the vicinity of a fluoropolymer producing factory. Sites with former fire-fighting activities are also potential *"hot spot"* areas. Concentrations encountered in drinking water remain fairly low on average. Typical concentrations are in the low ng/L range with the exception of highly contaminated areas, like in the Möhne and Ruhr area in Germany. The encountered concentrations in drinking water depend on the treatment technologies used to purify the water. Drinking water prepared with activated carbon or reverse osmosis will in general contain lower concentrations in tap water than in the raw water. However, the efficiency of water treatment depends much on the local boundary conditions.

C. Eschauzier • P. de Voogt
KWR Watercycle Research Institute, P.O. Box 1072, 3430 Nieuwegein, The Netherlands

Earth Surface Sciences, Institute for Biodiversity and Ecosystem Dynamics, Universiteit van Amsterdam, P.O. Box 94216, 1090 GE Amsterdam, The Netherlands

H.-J. Brauch • F.T. Lange (✉)
DVGW Water Technology Center, Karlsruher Str. 84, 76139 Karlsruhe, Germany
e-mail: frankthomas.lange@tzw.de

T.P. Knepper and F.T. Lange (eds.), *Polyfluorinated Chemicals and Transformation Products*, Hdb Env Chem (2012) 17: 73–102,
DOI 10.1007/978-3-642-21872-9_5,

Keywords Drinking water • Monitoring • Polyfluorinated chemicals • Water treatment

Contents

Abbreviations

6:2 FTS	6:2 Fluorotelomer sulfonate
AFFF	Aqueous film forming foam
ARW	Association of waterworks in the Rhine River basin
AWBR	Association of waterworks Lake Constance-Rhine
BAT	Best available technique
DWI	Drinking water inspectorate (UK)
EPA	Environmental protection agency (USA)
GAC	Granular activated carbon
HRIV	Health related indication values
LOQ	Limit of quantitation
PBSF	Perfluorobutane sulfonyl fluoride
PFBA	Perfluorobutanoic acid
PFBS	Perfluorobutane sulfonate
PFC	Polyfluorinated chemical
PFHpA	Perfluoroheptanoic acid
PFHxA	Perfluorohexanoic acid
PFHxS	Perfluorohexane sulfonate
PFNA	Perfluorononanoic acid
PFOA	Perfluorooctanoic acid
PFOS	Perfluorooctane sulfonate
PHA	Provisional health advisories
POP	Persistent organic pollutant
PTFE	Polytetrafluoroethylene

RIWA	Association of River Water Supply Companies
TZW	DVGW Water Technology Center (Technologiezentrum Wasser)
UBA	Federal Environment Agency (Germany)
WWTP	Wastewater treatment plant

1 Introduction

Polyfluorinated chemicals (PFCs; here: fluoroalkyl chemicals including perfluorinated chemicals as the special case where all hydrogens in the alkyl chain are substituted for fluorine), in particular short chain fluorinated alkyl sulfonates and carboxylates, are ubiquitously found in the environment. Their persistence and the bioaccumulative and toxic properties of some members of this compound class have instigated a considerable scientific, public and governmental concern and interest [1]. PFCs are found from the low ng/L to the low μg/L range in different types of environmental samples, such as surface waters [2–4], groundwater [5, 6], drinking water [7, 8], sea water [9, 10], sediments [11, 12], biota [13–15], food items [16] and blood serum [17]. This paper reviews the presence of polar PFCs in surface waters, groundwater and drinking water in Europe.

Although severe environmental concern arose not until the 1990s, the manufacture and processing of the diverse classes of fluorochemicals started about 60 years ago. The role they take in our everyday life has become increasingly important. They are used in a wide range of products and processes because of their unique properties. Differing surfactant properties of the various head groups and carbon skeleton chain lengths make that these surfactants are produced and used in many forms, for example for fluoropolymer synthesis and aqueous film forming foams (AFFFs). Furthermore, derivatives like esters and sulfonamides are used for leather, paper and textile finishing, as well as for impregnation of food packaging. It is the specific properties such as water, fat and dirt repellence, thermal and chemical stability, microbial inertness, and surface tension lowering that makes PFCs interesting for a multitude of commercial applications [18].

Recent actions taken by authorities in order to prevent further environmental contamination have led to several reductions in environmental emissions in the immediate past or near future. The voluntary initiative launched in 2006 by manufacturing industries to reduce emissions of perfluorooctanoic acid (PFOA) to the environment by 95% until 2010 (2000 as baseline year) is one example [19]. Although involved western industries aim at stopping PFOA emissions from products or facilities by 2015 [19], one should be aware that the phase-out of emissions does not entail global production stop. Recently, perfluorooctane sulfonate (PFOS) has been classified as a persistent organic pollutant (POP) by the Stockholm convention [20]. Also a restrictive regulation on the use of PFOS in Europe has been accepted by the European Parliament in 2006 [21]. According to the directive industries which cannot operate without PFOS are bound to use the best available techniques (BAT) to reduce emissions to the environment [21] and consumer products (semi-finished products or articles) may not contain more than 0.1 wt% of PFOS. The short-chain perfluorobutane sulfonyl fluoride (PBSF) and its derivatives were introduced by the *3 M Company* to replace the C8 homologues [22]. The C4 compounds are less bioaccumulative and toxic, but remain persistent in the environment.

Prevedouros et al. [23] distinguished two types of sources to the environment: direct and indirect sources.

Direct sources involve the use of consumer products (e.g., leaching from water and stain repellents), manufacture and use of PFC salts and fluoropolymers (such as polytetrafluoroethylene, PTFE) and especially the use of AFFFs (associated with high levels of non-branched and branched perfluorohexanoic acid (PFHxA), PFOA, perfluorohexane sulfonate (PFHxS), PFOS and 6:2 fluorotelomer sulfonate (6:2 FTS) [4, 5, 24]. In general, the actual discharge into the environment will occur via industrial or municipal wastewater treatment plants (WWTPs) [25–27], via direct emission to air, or through an AFFF [4, 5] or industrially contaminated area. In summary, known anthropogenic activities which can release significant quantities of PFCs are industrial WWTPs (depending on the activities), landfill leachate WWTPs [28, 29], (former) AFFF training areas [5] and (former) landfills. These "*hot spots*" have been related to elevated surface, ground- and drinking water contamination in several areas (see below).

Indirect emissions are caused by atmospheric degradation of precursor compounds. Atmospheric degradation of precursors is likely the major source of pollution in remote areas [30, 31]. Municipal WWTP effluents and infiltration of urban runoff and leaching piping [6, 32] are probably the major source of *diffuse pollution* to rivers and groundwater aquifers.

This paper aims at giving an overview of PFC concentrations found in European surface, ground- and drinking waters. Furthermore, an overview of characteristic sources of PFCs to the environment is given. Because peer-reviewed literature available on the presence and behavior of PFCs in European ground- and drinking water is still scarce some "grey literature" was also included, such as reports and websites of official institutions. Where necessary, data from outside Europe were also used to illustrate specific contamination examples for which no data exist in Europe.

2 PFC Concentrations in Surface Waters in Europe

A number of surface waters in Europe have been shown to contain PFCs as anthropogenic trace pollutants (Table 1). Concerning sources of surface water contamination, WWTPs play an important role. Municipal, industrial [26, 27, 56] and treated landfill leachate WWTP effluents [28, 29] have been proved to discharge PFCs and to increase environmental concentrations in rivers and also in groundwater aquifers (see Chap. 3. "PFC Concentrations in Groundwater"). One study [57] was able to correlate the mass-flow of PFOA and PFOS to the number of

Table 1 PFC in different European rivers (arithmetic mean values in ng/L; n.d. = not detected; <LOQ = below limit of quantification)

River or lake (sampling location)	Location	Country	# of samples (*n*)	Reference	Sampling year	PFBA	PFHxA	PFHpA	PFOA	PFNA	PFBS	PFHxS	PFOS
Dalälven		Sweden		Maclachlan 2007 [3]	2005		<LOQ	0.4	<LOQ	<LOQ			
Vindelälven		Sweden		Maclachlan 2007 [3]	2005		<LOQ	0.2	<LOQ	0.2			
Kalix Älv		Sweden		Maclachlan 2007 [3]	2005		<LOQ	0.3	<LOQ	<LOQ			
Örebro		Sweden	2	Lien 2006 [33]	2005				0.7				0.6
Lake Mjøsa		Norway	4	Kallenborn 2004 [34]	2003		1.2		6.8	0.2	<LOQ	0.1	0.3
Elbe		Germany		Maclachlan 2007 [3]	2005		15.4	2.7	7.6	0.3			
Elbe		Germany	10	Ahrens 2009 [35]	2007		3.4	1.4	7.6	0.7	2.3	1.0	1.6
Elbe	Hamburg	Germany		Ahrens 2009 [36]	2006	3.0	5.6	2.7	11.4	1.8	1.6	0.6	6.4
Elbe	Between Lauenburg an Hamburg	Germany		Ahrens 2009 [36]	2006	2.2	4.4	2.9	8.0	1.7	1.1	0.4	5.5
Aare	Felsenau	Germany		AWBR 2008 [94]	2007		<LOQ	<LOQ	2	<LOQ	2	2	8
Roter Main		Germany	4	Weremiuk 2006 [95]	2005				15				27
Rhine	Mainz										3.8		
Rhine	Mainz										1.2		
Rhine	Au-Lustenau (before Lake Constance)	Austria		AWBR 2008 [94]	2007		<LOQ	<LOQ	<LOQ	<LOQ	<LOQ	<LOQ	4
Rhine	Basel-Birsfelden	Germany		AWBR 2008 [94]	2007		<LOQ	<LOQ	1	<LOQ	2	1	6
Rhine	Basel-Birsfelden	Germany		AWBR 2009 [96]	2008		<LOQ	<LOQ	<LOQ	<LOQ	<LOQ	<LOQ	8
Rhine	Basel-Birsfelden	Germany		AWBR 2010 [97]	2009		<LOQ	<LOQ	<LOQ	<LOQ	<LOQ	<LOQ	8
Rhine	Karlsruhe	Germany		AWBR 2008 [94]	2007		1	1	2	<LOQ	14	4	13
Rhine	Karlsruhe	Germany		AWBR 2009 [96]	2008		1	<LOQ	<LOQ	<LOQ	4	<LOQ	10
Rhine	Karlsruhe	Germany		AWBR 2010 [97]	2009		<LOQ	<LOQ	<LOQ	<LOQ	<LOQ	<LOQ	11
Rhine	Mainz	Germany	14	ARW 2009 [43]	2008	2.9	1.1	<LOQ	2.9	<LOQ	3.8	2.3	11.9
Rhine	Mainz	Germany	13	ARW 2008 [98]	2007		1.0	<LOQ	2.5	<LOQ	4.9	2.2	8.9
Rhine	Köln	Germany	14	ARW 2009 [43]	2008	0.2	1.4	0.4	2.6	<LOQ	3.4	2.5	10.9
Rhine	Köln	Germany	13	ARW 2008 [98]	2007		1.0	0.61	3.0	<LOQ	3.6	1.7	8.3
Rhine	After Leverkusen	Germany		Moller 2009 [42]	2008	116 ± 40					45.4 ± 30		
Rhine	Before Leverkusen	Germany		Moller 2009 [42]	2008	2.3					3.8		

(continued)

Table 1 (continued)

River or lake (sampling location)	Location	Country	# of samples (*n*)	Reference	Sampling year	PFBA	PFHxA	PFHpA	PFOA	PFNA	PFBS	PFHxS	PFOS
Rhine	Düsseldorf-Flehe	Germany	14	ARW 2009 [43]	2008	90.4	1.3	0.4	3.1	<LOQ	70.7	2.5	13.4
Rhine	Düsseldorf-Flehe	Germany	13	ARW 2008 [98]	2007		1.1	0.6	3.0	<LOQ	39.7	1.5	9.2
Rhine	Düsseldorf/Duisburg	Germany		Lange 2007 [58]	2006						81/48		
Rhine	Wesel	Germany		Loos 2009 [52]									32
Rhine		Germany		Maclachlan 2007 [3]	2006		18.2	1.8	11.6	0.6			
Rhine		Germany		Maclachlan 2007 [3]	2006 (Feb)		3.3	3.3	12.3	1.5			
Rhine	Lobith	Netherlands	13	Riwa 2010 [46]	2009	31.9	1.2	1.2	4.3	<LOQ	21.4	2.1	14.5
Rhine	Lobith	Netherlands	9	Riwa 2009 [47]	2008	69.6	2.1	1.3	3.8		47	2.7	24.2
Rhine	Lobith	Netherlands		de Voogt 2006 [48]				10.5	55.5	27.1		3.4	35.4
Rhine	Lekkanaal Nieuwegein	Netherlands	4	Riwa 2010 [46]	2009				5.4				8.4
Rhine	Lekkanaal Nieuwegein	Netherlands	4	Riwa 2009 [47]	2008				n.d.				8.4
Rhine	Lekkanaal Nieuwegein	Netherlands	13	Riwa 2008 [49]	2007				n.d.				8.6
Rhine	Lekkanaal Nieuwegein	Netherlands	13	Riwa 2007 [50]	2006				6.3				13
Rhine	Amsterdam Rijnkanaal	Netherlands	4	Riwa 2009 [47]	2008				6.1				8.9
Rhine	Amsterdam Rijnkanaal	Netherlands	13	Riwa 2007 [50]	2006				8.1				14.1
Meuse		Netherlands	5	de Voogt 2006 [48]				<LOQ	31.2	<LOQ	<LOQ	<LOQ	28.5
Across the Netherlands		Netherlands	21	Kwadijk 2010 [51]					6.5–43		6.4–290		4.7–32
Scheldt		Netherlands		Loos 2009 [52]					73				110
Scheldt		Netherlands	2	de Voogt 2006 [48]	2006			<LOQ	27	<LOQ	<LOQ	19	<LOQ
Scheldt		Belgium		Loos 2009 [52]					88				154
Po		Italy		MacLachlan 2007 [3]			19	6.6	200	1.5			
Po		Italy	11	Loos 2008 [64]	2007			2.4	89	1.8			6.1
Lago Magiore		Italy	8	Loos 2007 [82]					2.4	0.6			7.8
Lago Trassimano		Italy		de Voogt 2006 [48]	2006			3.5	3.3	1.7	3.4	1.8	3.1
Danube		Austria		Loos 2009 [52]					25				
Danube		Austria	3	Clara 2009 [99]			n.d.	n.d.	18	n.d.			<LOQ
Schwechat		Austria	3	Clara 2009 [99]			3.0	1.1	3.4	<LOQ			18.7
Liesing		Austria	3	Clara 2009 [99]			2.0	1.9	10.3	<LOQ			15.3

Rhone	France		Loos 2009 [52]				116				
Seine	France		Loos 2009 [52]								97
Seine	France		Maclachlan 2007 [3]	2005	13.3	3.7	8.9	1.3			
Loire	France		Maclachlan 2007 [3]	2005	3.4	0.9	3.4	0.4			
Wyre	UK		Loos 2009 [52]				100				
Svern	UK		Loos 2009 [52]								238
Krka	Slovenia		Loos 2009 [52]								1,371
Catalonia (Northern Spain)	Spain	4	Ericson 2008 [8]		<LOQ	2.1	9.4	0.5	<LOQ	0.5	3.3
Glatt	Switzerland		Huset 2008 [100]		n.d.	1.4	7.4	n.d.	4.3	12.3	49

Values below LOQ were taken as 0 in the calculation of the mean values.

inhabitants in a watershed indicating that municipal WWTPs certainly contribute to PFC discharges into the environment. However, beyond a discharge threshold of PFOA of 0.5 tons per year this relation did not hold anymore. An increased influence of point sources was expected to be an explanation. PFC concentrations in the European rivers are discussed in rough geographical order from North to South.

2.1 *Northern Europe*

The available reports about PFCs in Nordic surface waters present relatively low concentrations in comparison with the rest of Europe (Table 1). The low population density and fewer industrial activities in Scandinavian countries compared to central Europe could explain the lower concentrations found in the North of Europe. One study, in which Norwegian lake water was analyzed ($n = 4$), found low concentrations of PFCs. PFOA was measured at the highest concentration of 8.2 ng/L and the PFOS concentration was 0.48 ng/L [34]. In Swedish rivers and lakes McLachlan et al. [3] reported concentrations below 0.36 ng/L for perfluoroheptanoic acid (PFHpA), PFOA, and PFOS, and Lien et al. [33] reported average PFOA and PFOS concentrations of 1.7 and 1.9 ng/L respectively around Örebro (see Table 1).

2.2 *Central Europe*

The Rhone, Rhine, Danube, and Po rivers have the highest discharges of the European rivers considered (between 810 and 2,200 m^3/s at sites sampled for PFC analyses) and in part also high PFC concentrations, thus generating a considerable mass flux of PFC even at low water contamination levels.

Concentrations of PFCs in the Rhine River were monitored extensively in Germany and in the Netherlands as can be seen in Table 1. Background values for most PFCs were in the low ng/L range, i.e. <10 ng/L. Mainly, perfluorobutanoic acid (PFBA) and perfluorobutane sulfonate (PFBS) were found in high concentrations in the Rhine, and PFOA and PFOS in other rivers. Other short chain PFCs (<C9) were found to be present but often in low concentrations.

Many of the measurements in the Rhine catchment area were done by AWBR (Association of Waterworks Lake Constance-Rhine), ARW (Association of Waterworks in the Rhine River Basin), and RIWA (Association of River Water Supply Companies). The concentrations reported from the different locations on the Rhine River are discussed below from the upper Rhine to the lower Rhine including tributaries. For clarity purposes Figure 1 shows the catchment area of the Rhine river with important confluents.

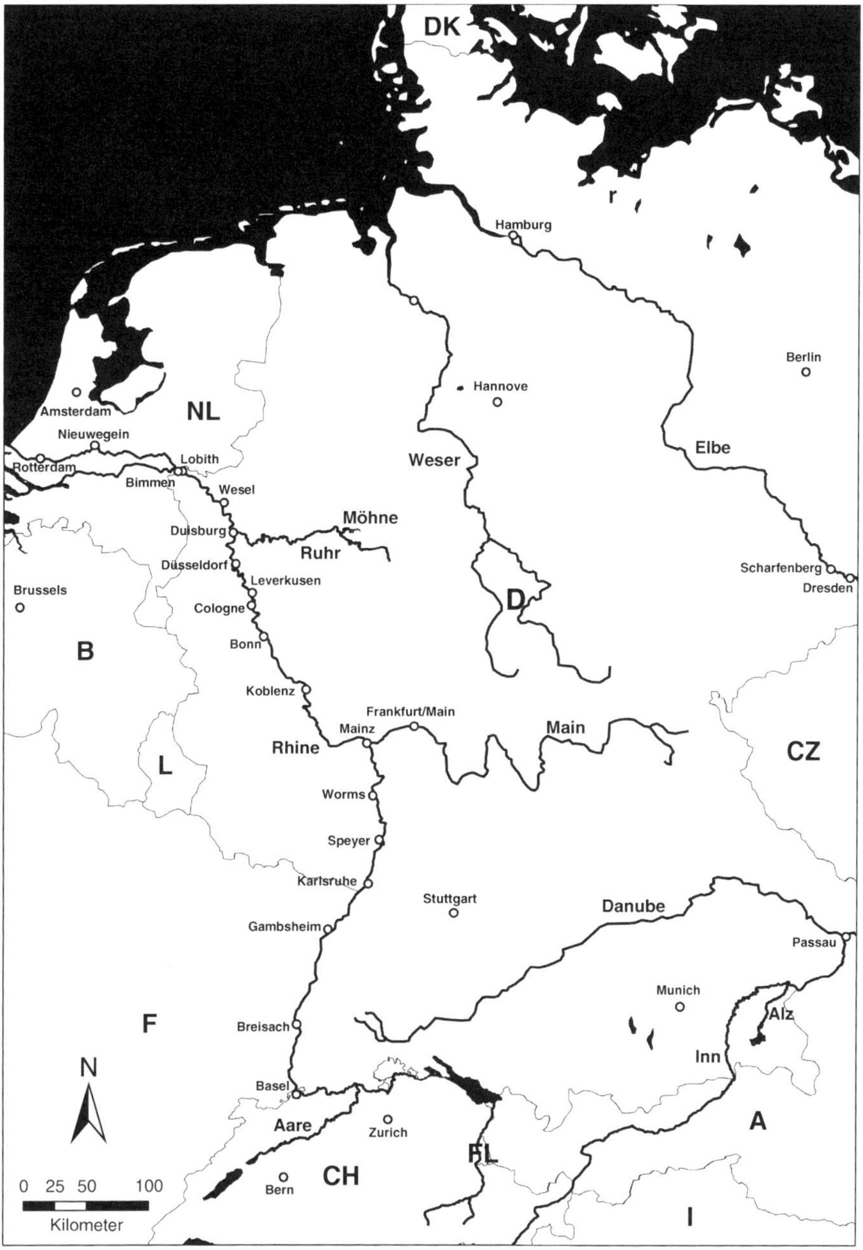

Fig. 1 Catchment area of the river Rhine and selected other rivers mentioned in this chapter

In autumn 2006, a maximum PFBS concentration of 2,900 ng/L was measured in the upper Rhine within a period of about two weeks [58]. This high level was a consequence of a contamination in the Aare River in Switzerland before the confluence with the Rhine due to a still unknown temporary emission into the Aare catchment area.

At the Dutch–German border at Lobith high concentrations of PFBA and PFBS were observed in 2008 with average concentrations (monthly grab samples during one year) of 70 and 47 ng/L respectively [59]. It was proved that one WWTP discharging industrial wastewater in the lower Rhine in Germany around Leverkusen was responsible for an increase of PFBA and PFBS concentrations from the low ng/L range (<5 ng/L) to 117 ± 40 and 45 ± 30 ng/L after the WWTP [42]. The ARW [43], which reported on different PFC concentrations at Mainz, Köln and Düsseldorf-Flehe, observed the same increase in concentrations (Figs. 2 and 3). Concentrations of PFBA and PFBS in Mainz and Köln were low throughout the year 2008 (Table 1) whereas mean concentrations in Düsseldorf-Flehe were 90 and 71 ng/L for PFBA and PFBS, respectively. Earlier analysis in spring 2006 by Skutlarek et al. [2] found a PFBS concentration of 15 ng/L in the lower Rhine

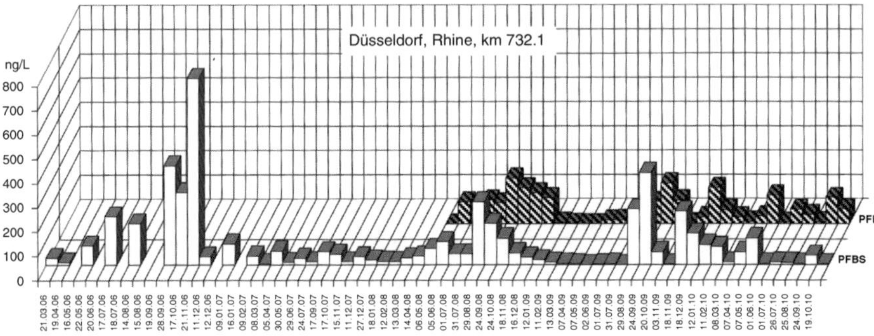

Fig. 2 PFBA and PFBS concentrations (in ng/L) in the Rhine River at Düsseldorf (km 732.1) from 2006 to 2010; data from [43, 58] and complemented with recent data

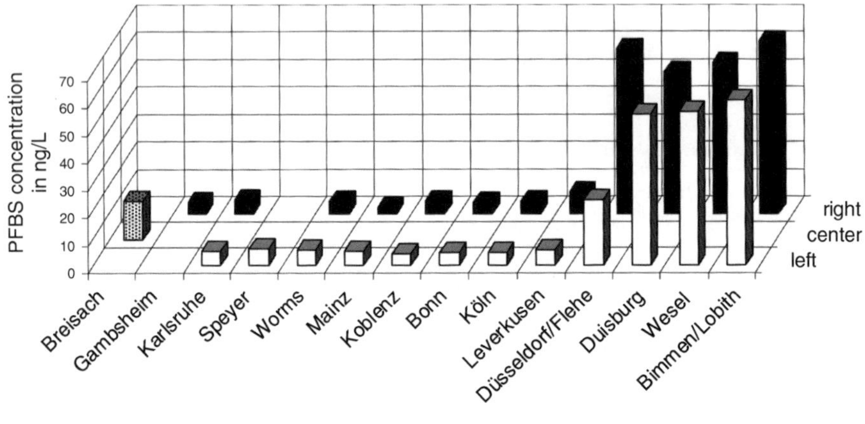

Fig. 3 PFBS concentrations (in ng/L) across the Rhine River (left and right bank and centre at May 8, 2006 [58]

around Duisburg, which is situated downstream of Leverkusen. This level was within the typical range of PFBS concentrations in the Rhine during the sampling period and much lower compared to the concentrations found by the ARW [43] and Moeller et al. [60] in 2008. Overall, the concentration of PFBS and PFBA in the lower river Rhine seems to be relatively high compared to other PFCs. Especially for PFBS a further increase can be expected as the short chain PFCs will be increasingly used in the future.

The fact that the spontaneous PFBS concentration increase downstream of Leverkusen is caused by a point source can be clearly identified by the distribution of PFBS in the cross-section of the Rhine River (Fig. 3), which indicates an emission at the right bank of the river and a complete mixing across the section further downstream until the German/Dutch border at Bimmen/Lobith.

The influence of individual point sources was also indirectly shown at another sampling site at the Rhine River in Cologne (Fig. 4a), where the correlation between the PFOA and PFOS concentration and the reciprocal river discharge was not significant due to the influence of numerous point sources. This is contrary to what was observed in the Elbe River (see Fig. 4b).

In North Rhine-Westphalia, Germany, in May 2006, the application of an illegally contaminated so-called soil improver on agricultural land was detected and caused the release of large quantities of PFCs into the Möhne catchment area, a tributary of the Ruhr River. The Ruhr River, which confluences with the Rhine River became highly contaminated mainly with PFOA and some other PFCs [2]. Sampling in the Rhine downstream of the Ruhr and Rhine confluence showed low PFC concentrations ($\sum$PFC = 41 ng/L), whereas in samples collected from the Möhne River very high concentrations around Heidberg ($\sum$PFC = 4,385 ng/L) and around Bestwig ($\sum$PFC = 4,268 ng/L) were observed. Monitoring at regular time intervals by the local authorities since 2006 and a sampling campaign in 2008 showed a maximum total PFC concentration in the Möhne just upstream of the confluence with the river Ruhr of 309 ng/L (PFBA, PFPeA, PFHxA and PFOA dominated) [60]. This is considerably lower than the maximum concentrations Skutlarek et al. reported in 2006. Apparently, PFC concentrations in surface waters in the river Möhne catchment are steadily decreasing with time.

Further downstream in the Netherlands, in the Lekkanaal, average (n = 30) annual concentrations of PFOA and PFOS were below 30 ng/L for each compound in the period 2006–2009 [46, 47, 49, 50] (Fig. 5). Linear-regression analysis shows a significant decreasing PFOS concentration trend over the last three years ($P = 0.0198$; despite a low r^2 of 0.179), which is probably due to the PFOS production stop in 2002.

Kwadijk et al. [51] who analyzed surface water samples (n = 21) collected across the Netherlands observed concentrations between 6.4 and 290 ng/L for PFBS with the highest concentration measured in the Rhine River at Lobith. This corresponds fairly well to the measurements performed by AWBR, RIWA and ARW (see above). PFOA was measured between 6.5 and 43 ng/L and PFOS between 4.7 and 32 ng/L. Measurements performed for the PERFORCE project [48] resulted in average concentrations of 19 and 28.3 ng/L for PFOA and PFOS,

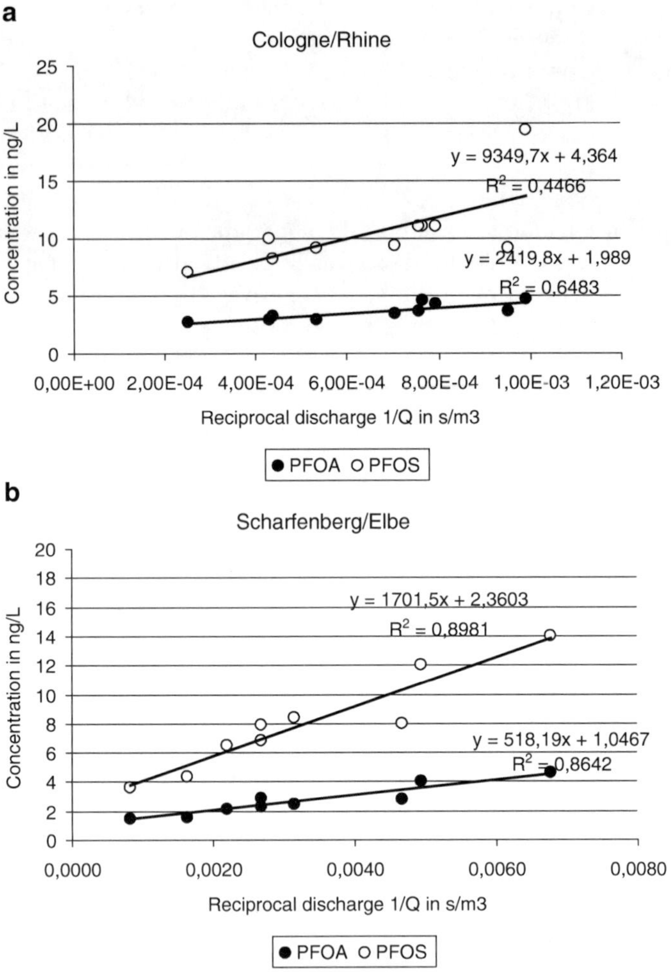

Fig. 4 (**a**) PFOA and PFOS concentrations (in ng/L) in the Rhine River at Cologne (km 684, left bank) in 2006 and (**b**) in the Elbe river at Scharfenberg (km 76, right bank) in 2006 [61]

respectively, in the Dutch part of the Rhine River, which corresponds to the findings of Kwadijk et al. [51]. However, locations were not specified in this report. Other PFCs measured in the PERFORCE project were reported to be below the limit of quantitation (LOQ), which at the time of analysis (2005) were still high (e.g., 23 ng/L for PFBS).

In the Netherlands a sprinkler installation at the Amsterdam Schiphol airport accidently released large amounts of AFFF containing PFCs in July 2008. The contaminated water was collected, diluted and discharged into a WWTP in the area, which discharges its effluent into the surrounding ditches and canals. A following monitoring campaign conducted by the Dutch government showed peak

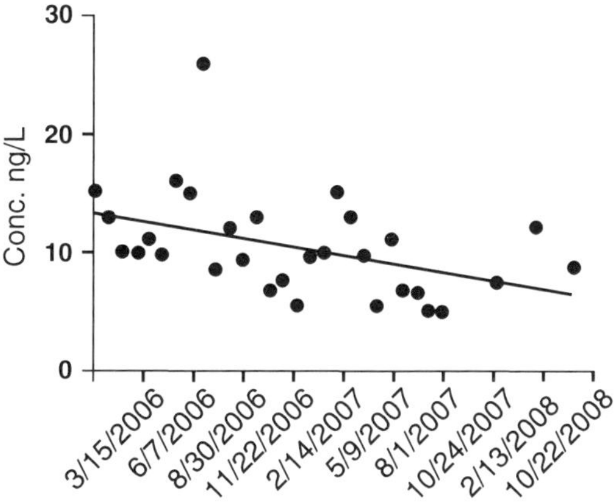

Fig. 5 Concentration of PFOS (in ng/L) in the Rhine River at Lekkanaal (Nieuwegein, the Netherlands) sampled in the period from 2006 to 2008 (based on [47, 49, 50])

concentrations in the North Sea canal (location Halfweg) of PFOS of 1,300 ng/L which decreased to 100 ng/L after two months. The PFC profiles observed in the surrounding surface waters showed a large contribution of PFOS, PFHxS and PFBS to $\sum$PFC, which is typical for AFFF contaminations [62].

In the Elbe at Scharfenberg, downstream of the city of Dresden, Germany, the concentrations of PFOA and PFOS (sampled in 2006) correlate fairly well with the reciprocal river discharge (see Fig. 4b). This correlation is a clear indication that the relatively low concentrations observed in the river are dominated by diffuse sources [61]. Two further publications report on the concentrations of PFCs along the Elbe River [35, 36]. The mass flow of PFCs in the Elbe River is rather low compared to the Rhine and Po Rivers as a result of the lower concentrations and the lower river discharge ($\pm$300 m^3/s). Predominating substances measured in 2007 were (mean concentrations measured along the Elbe River) PFHxA with 3.4 ng/L, PFOA with 7.6 ng/L, PFBS with 2.3 ng/L and PFOS with 1.6 ng/L [35]. A subsequent sampling campaign performed a year later revealed the same predominating substances in lower concentrations except for PFOS, which was higher than in 2006. Its mean concentration was 6.4 ng/L around Hamburg [36]. Furthermore from Fig. 4b, which represents a situation with predominating diffuse PFCs inputs, PFOS to PFOA ratio of $\approx$3:1 can be deduced, at least for measurements in Germany in 2006. Larger deviations from this rule of thumb indicate an important contribution of point sources to PFC pollution.

Such a situation is the high concentrations found in the river Alz in Germany in 2007 in the vicinity of a fluoropolymer manufacturing facility [63]. Surface water samples [41] ($n = 20$) showed a maximum total PFC concentration of 8,000 ng/L from which 7,500 ng/L were from PFOA. Downstream, in the Inn and Danube concentrations of 100 and 50 ng/L PFOA were measured, respectively. For groundwater and drinking water concentrations, see the corresponding sections. Loos et al. [52] found high concentrations of PFOA on one occasion in the Krka River in Slovenia (up to 1,400 ng/L). Although the concentrations encountered seemed high, the flow of the river was relatively small (50 m^3/s) compared to the main European rivers. In the Seine River in France, PFOA [3] and PFOS [52] concentrations of 8.9 and 97 ng/L were measured, respectively.

The mass discharge of PFC into European rivers was shown to correlate with the population (below a threshold of 0.5 tons per year) of the catchment and thus partly explain the higher concentrations encountered in populated areas [57]. The measured concentrations are usually highly variable in space and time, such as measured in the Rhine River, making data verification difficult if not impossible.

2.3 Southern Europe

Several studies reported high concentrations of PFOA in the Po River, Italy. Loos et al. [64] observed a mean concentration of PFOA of 89 ng/L with a maximum of 337 ng/L and McLachlan et al. [3] reported a mean concentration of 200 ng/L. Recent sampling in the Po watershed showed that several fluoropolymer manufacturing plants located around the city of Alessandria and further downstream around the confluence of the Po and Bormida Rivers are the main sources of PFC pollution [65]. In Catalonia, Spain, PFC concentrations found in the Ebro, Cortiella and Francoli rivers were highest for PFOA (24.9 ng/L) and PFOS (5.9 ng/L), both in the river Francoli [8].

2.4 Western Europe (United Kingdom)

Following an explosion at the Buncefield oil depot in December 2005, considerable amounts of fire fighting foams containing PFOS were released to the surrounding surface waters next to the hazardous site. Monitoring data from the Buncefield area, reported by the Environment Agency in the United Kingdom (UK) [66] show relatively low continuous concentrations of PFOS in surface waters in the vicinity of the depot area over time after the accident. Groundwater in the immediate vicinity of the explosion site appeared more heavily polluted with PFCs. In 2007, an extensive monitoring program was started to assess 19 different drinking water treatment locations (raw water, some treatment steps and drinking water) throughout England. Locations selected for sampling were typically areas in the vicinity of an airstrip, industrial area, or known polluted sites (sewage discharge, Buncefield). This survey reported maximum concentrations of 370 and <11 ng/L for PFOA and PFOS, respectively, in surface waters [67].

2.5 Eastern Europe

A study in Poland reported low concentrations of PFCs in surface waters in the North of Poland and the Baltic Sea [37]. In Southern Poland, one sampling location was reported to have average concentrations of 152, 106 and 31 ng/L for PFOS, PFHxS and PFHxA, respectively. At the other locations PFCs were measured below 18 ng/L, including PFBS and PFOA.

3 PFC Concentrations in Groundwater

Little information is available on background concentrations of PFCs in European groundwater or in groundwater from other parts of the world. However, from "grey" literature it can be concluded that typical sources of groundwater contamination are contaminated fertilizers (soil improver or sewage sludge), percolating AFFFs, infiltrating surface waters (e.g., bank filtrate), and possibly, leaching landfills or diffuse urban pollution (leaking sewers and surface runoff). Since the remediation of contaminated soils is expensive, and generally, hardly any remediation of the contaminated sites is performed, leaching of PFCs into the environment for a long period of time is likely and should be taken seriously regarding the extent of the contamination at sites severely polluted with PFCs. Due to the scarcity of PFC data in groundwater aquifers, some examples from outside of Europe are also compiled in this section in order to describe the relevant input pathways. Groundwater treatment facilities often have a less pronounced multi-barrier treatment system compared to surface water treatment, and adsorption is not a powerful removal mechanism for short chained PFCs (see "Sorption and Leaching Behavior of Perfluorinated Compounds in Soil" of this book). Therefore, PFCs present in groundwater can travel relatively easily through the pertaining water treatment systems and may thus give rise to human exposure.

3.1 PFCs in Groundwater at Severely Polluted Sites

In Bavaria, Germany, since 2007, groundwater samples ($n = 97$) have been analyzed from the near vicinity of the industrial area Gendorf (around the village of Emmerting), which was also known as a surface water hot spot of the small river Alz [63]. In this area groundwater contamination with PFOA (especially the Alztal aquifer) was up to 7,000 ng/L [38, 39]. The groundwater pollution was revealed through the presence of PFCs in the drinking water of the region, and aquifers used as source water were found to contain up to 4,300 ng/L of PFOA in the Inn-Salzachgruppe (see also drinking water section). This contamination is known to stem from the emission of PFOA used as an emulsifier in the production of fluoropolymers. Nowadays, PFOA is substituted by an alternative PFC. However, the identity of this substitute is confidential information.

Another contamination in German groundwater was detected in 2009 in the catchment of a waterworks of the RheinEnergie AG near Cologne. The source of the contamination was identified to be a fire brigade training area and a site contaminated with AFFFs. ∑PFC reached levels up to 4,000 ng/L with PFOS and PFHxS prevailing. The PFC pattern was as follows: PFOS (>80%), PFHxS (8–12%), PFHxA (3–4%), PFBS (1–3%) and PFOA (1–2%) [40]. PFC concentrations in the drinking water were reduced by blending with clean raw water and subsequent removal using granular activated carbon (GAC) filtration.

In North Rhine-Westphalia, Germany, the local water supplier in Lippstadt closed down the groundwater waterworks Eikeloh in October 2006 when the sum of PFOS and PFOA exceeded 500 ng/L. After the installation of GAC filters in February 2007, the waterworks could be re-opened. It appeared that the source of contamination was the application of soil improver [40].

Surface water influence on groundwater quality was also observed in the neighborhood of the creek Rheder Bach in North Rhine-Westphalia, which is contaminated with PFCs by emissions from the local municipal WWTP receiving industrial wastewaters of two PFC emitting companies. In the creek, concentrations of 1,100 ng/L for PFOA and 360 ng/L for PFOS were measured [44]. In the groundwater, the sum of PFOA and PFOS was 279 ng/L, close to the guidance value of 300 ng/L for drinking water as given in a recommendation of the German Federal Environment Agency (UBA) [45].

Another well documented case study (not in Europe) is the PFC contamination around a landfill site where production waste from a perfluorochemicals manufacturing plant was dumped. In 2004, it appeared that PFCs were present in groundwater in local municipal and private wells in Oakdale, USA, (situated south of one of the landfills) [53, 54] (Fig. 6) and in local tap water at concentrations above US Environmental Protection Agency (EPA)'s Provisional Health Advisories (PHA) levels (see drinking water section). A GAC treatment plant was installed and filtration began at the end of 2006 in order to remove PFCs from the drinking water [55]. Given low groundwater velocity in general, a contaminated site will cause problems by dispersing slowly and remaining present for possibly tens of years (e.g., [5]). Figure 6 shows that concentrations of different PFCs only slightly decrease over a time range of several years. Sources still existed for the given data and are currently being remediated.

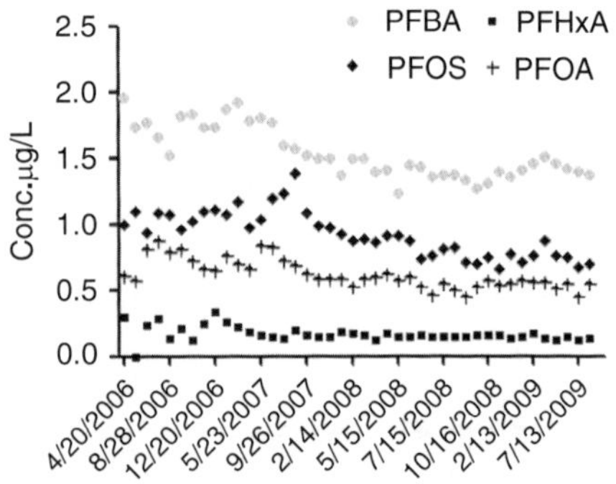

Fig. 6 Concentrations of PFC in a groundwater well in Oakdale used for drinking water purposes, USA (courtesy of the Minnesota Department of Health, personal communication with Kolstad Chad)

The release of fire-fighting foams due to fires, accidental releases, or fire-fighting trainings is known to cause contaminations of groundwater in often high concentrations [68–71]. Moody and co-workers [5] found rather high concentrations of four PFCs in ten different groundwater wells at an Air Force base in Michigan, USA. Maximum concentrations amounted to 120,000 ng/L for PFHxS, 110,000 ng/L for PFOS, 20,000 ng/L for PFHxA and 105,000 ng/L for PFOA.

Recent monitoring in the UK also revealed the presence of PFOA and PFOS in groundwater used for drinking water production. The source of this contamination was either pollution incidents (e.g., Buncefield explosion) or the vicinity of a local source such as an airstrip [67, 72]. Maximum PFOA and PFOS concentrations found in groundwater (i.e. influent of the drinking water treatment station) in this monitoring campaign were 230 and 152 ng/L, respectively.

Another contaminated site in the UK is the Jersey airport, where the "Airport Fire and Rescue Service" released significant quantities of AFFFs to the environment by fire-fighting trainings. The highest concentration of PFOS measured was 98,000 ng/L, however, concentrations up to 10,000 ng/L could still be measured in 2009 [72].

The analysis of landfill effluents collected in Finland and Norway resulted in a maximum concentration observed for ΣPFC of 1,537 ng/L [34]. In landfill effluents from 22 sites in Germany, a maximum concentration of ΣPFC of 13,000 ng/L was observed [29]. Although effluents of modern landfills are often collected and treated nowadays, many former landfills leach percolate water to groundwater aquifers and are a potential source of PFCs to drinking water wells. It might be reasonable to assume that the concentrations leached into the environment would have been in the same order of magnitude as encountered in collected leachate.

3.2 Monitoring Campaigns for PFCs in Groundwater

In 2006, in the state of Baden-Württemberg, Germany, 46 selected groundwater wells with potential PFC contamination were analyzed [73]. These wells were selected either due to a known direct or indirect impact of wastewater, e.g., from a sewage treatment plant site, due to known leakages in the sewer system, or due to surface water infiltration. Additional wells were chosen, which were located near sites where PFCs had been applied, such as paper finishing and electroplating plants. Other samples were taken from wells situated downstream of landfills or from sites where in the past there had been a major fire or regular fire-fighting trainings, i.e. at an industrial site and at a military airbase. In spite of the expected pollution, at approximately 80% of the sites selected ΣPFC (18 compounds) was below 50 ng/L. Therefore, it was concluded that there is no significant spatially conclusive and comprehensive contamination of groundwater in the state of Baden-Württemberg. The highest concentration was measured at a groundwater well close to the Rhine River, where a PFBS concentration of 2.5 μg/L was analyzed.

This could be understood in terms of the high temporal PFBS concentration in the upper Rhine valley at the time of sampling (see Chap. 2 “PFC Concentrations in Surface Waters in Europe”).

The analysis of 51 different groundwater samples in Bavaria, Germany, (excluding the Gendorf area mentioned above) in 2007 showed that at 13 sites PFCs were found. PFOA and PFOS concentrations ranged between 0.6 and 4.1 ng/L and <1–20 ng/L, respectively. Groundwater contamination was mainly associated by infiltration of river water for drinking water production [41].

The results from a small sampling campaign in Dutch groundwater used for drinking water production showed the presence of PFOA at 68 and 44 ng/L at two out of five sites sampled. At one site a concentration of PFNA of 14 ng/L was observed. It has to be noted that LOQs in this study where relatively high, i.e. in the 10–20 ng/L range [74].

To the best of our knowledge, the leaching of surface runoff and from sewer pipes has not been studied in Europe. One Japanese paper reports on the contamination of groundwater in the city of Tokyo [6]. PFHpA, PFOA, PFNA and PFOS were present in the following concentrations ranges: <0.1–20, 0.47–60, 0.1–94 ng/L and 0.28–133 ng/L, respectively. Surface runoff, wastewater leaching from sewer pipes, and in one sample infiltrating river water appeared to be the sources of the contaminations. This could be denoted as diffuse urban pollution.

4 PFCs in Drinking Water

4.1 Occurrence of PFCs in Drinking Water

Low levels of PFCs are regularly found in drinking waters across Europe. The relationship between elevated surface water or groundwater concentrations of PFCs on the one hand and drinking water concentrations of PFCs on the other was established in several papers and research programs [61, 75]. Drinking water from polluted areas, especially near airstrips, where spills or continuous emissions had occurred, contains elevated PFC concentrations.

Exposure assessment studies have concluded that both food and drinking water can be major exposure pathways to humans [16, 76]. It was also shown that contaminated drinking water yields higher blood plasma concentrations of PFOA in humans [77–79]. The consumption of drinking water was estimated to give <0.5% and 16% of the total human exposure to PFOS and PFOA, respectively [80]. However, data used for this assessment were limited in the concentrations of specific dietary items available for the assessment.

Concentrations of individual PFCs have been determined in drinking water in several studies. Statistical evaluation of 121 drinking water samples from 99 different origins in Germany and Switzerland [81] demonstrated that a number of analyzed polar to medium polar PFCs were frequently present in drinking water samples, even if highly contaminated areas were excluded (Fig. 7).

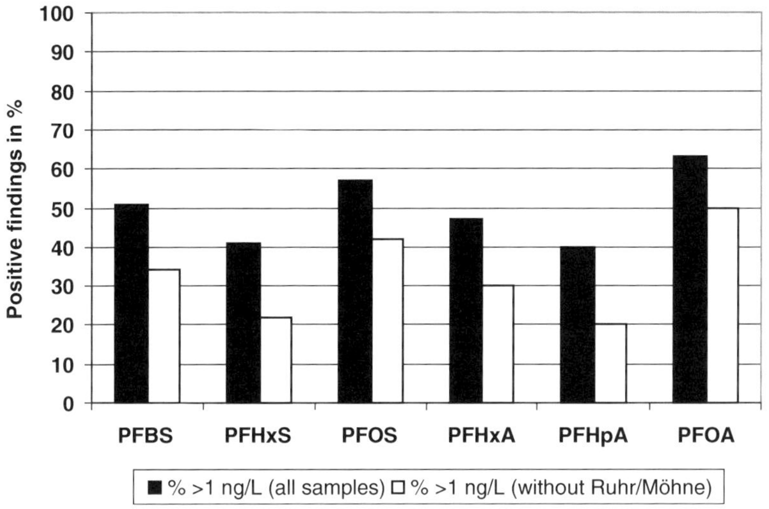

Fig. 7 Percentage of positive samples including and excluding the heavily contaminated Ruhr/Möhne area (non published data representation by F. T. Lange, TZW)

This finding reflects that PFCs are often present in drinking waters at very low levels and that the contaminated areas do not necessarily contribute to a large extent to the number of positive findings. This can be explained in part by the low LOQs reached nowadays by the analytical methods applied.

It was also observed that the severely contaminated sites do not contribute substantially to the median concentration of all 121 samples. This can be seen when comparing Fig. 8a and b. Figure 8a depicts the median concentration of the 121 measured samples with the outliers (mainly the contaminated sites) above the 90th percentile. Upon removing the values related to contaminated sites the median concentrations do not change much (Fig. 8b)). This indicates that in the majority of the locations sampled measured concentrations are low and that only in few cases guideline values locally can be exceeded.

A study in Catalonia, Spain, showed the presence of PFCs in tap water with maximum concentrations of 57, 69 and 58 ng/L for PFOA, PFBS and PFOS, respectively. Concentrations of other PFCs were below 10 ng/L (PFHxA, PFHpA, PFNA, PFDA, PFUnDA, PFDoDA, PFTDA, PFHxS, PFOS, and PFOSA) [7]. Another study measured tap water concentrations in Sweden near Örebro and found concentrations of 1.3 ng/L PFOA and 0.3 and 0.8 ng/L PFOS [33]. Loos et al. [82] found several PFC in tap water in the vicinity of Lake Maggiore in Italy. Only PFOA (2.4 ng/L) and PFOS (8.1 ng/L) were found in concentrations above 1 ng/L (PFBA and PFBS were not measured). Surrounding surface waters contained comparable PFC concentrations (see above) indicating that the water treatment used did not efficiently remove the PFCs. In a Belgian study, [83] it was observed that in tap water samples from three different communities in Flanders (Antwerp, Waasland and Gent; with $n = 4$), the median concentration of PFOS (3.4 ng/L) was

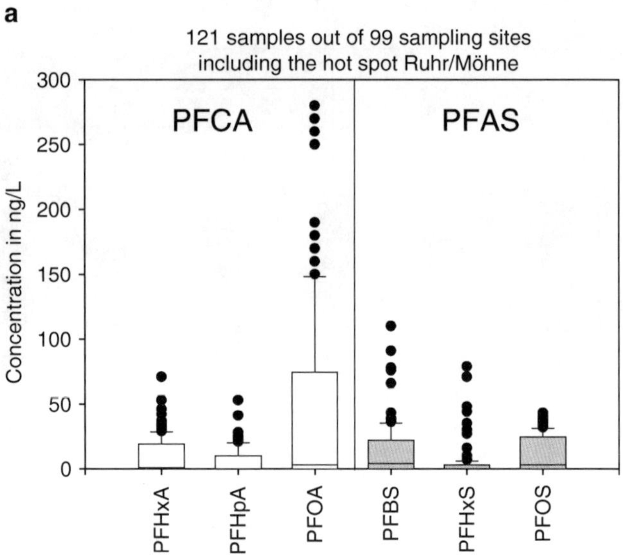

Fig. 8 Boxplots of PFC concentrations in drinking water (data of 2006) (**a**) including known hot spot samples and (**b**) excluding known hot spot samples; *black dots* represent outliers (mostly polluted sites), *line* within the box represents 50th percentile (median); box delimits 25th and 75th percentiles; *bars* indicate 10th and 90th percentiles; results <1 ng/L were taken as 0 ng/L in the calculation [61]

the highest of the PFCs analyzed, followed by PFOA and PFHxS (both 1.1 ng/L). The other PFCs (PFBA, PFHxA, PFNA and PFBS) analyzed were invariably below 1 ng/L, except for PFHxA for which a LOD of 1.8 ng/L was reported. One recent study in Norway reported concentrations of PFHxA, PFOA, PFHxS and PFOS of 0.36, 1.45, 0.11 and 0.20 ng/L, respectively [84].

The concentrations levels mentioned in the previous paragraph are regarded as low. Drinking water which is produced in the vicinity of a PFC-contaminated area has often higher concentrations compared to background areas. For example, the drinking water levels from waterworks situated in the Ruhr catchment area, which have been monitored closely since the detection of a high PFOA contamination in 2006 [2] following the application of a contaminated soil improver to agricultural land (see Chap. 2. PFC Concentrations in Surface Waters in Europe, Subchap. 2.2 "Central Europe"), have amounted up to levels sometimes above the precautionary value of 100 ng/L recommended for the sum of PFOA and PFOS concentrations. Timelines (since 2006) of PFOA and PFOS concentrations as well as for their combined concentration can be retrieved from [85].

In Southern Germany another area is known where environmental emissions of PFOA caused drinking water contamination. In the Altötting District (Bavaria), drinking water has been (and still is) monitored for PFCs from 2006 to 2009 [86] following discharges from a fluoropolymer factory using PFOA in the production process. Concentrations of PFOA between the LOD (1 ng/L) and 410 ng/L were reported. PFOS was not detected above 4 ng/L in these regions. At three locations in the Altötting area, a consistent increase between 2006 and May 2009 (Fig. 9) was observed. At several occasions the recommended health based orienting value for drinking water of 300 ng/L for the sum of PFOA and PFOS [87] was exceeded. The variation in the mixing of the different waters obtained from the different pumping stations possibly causes the temporal increase in concentrations at high level tank Vogled seen in Fig. 9 around April 2009. In November 2009, activated carbon filters were installed in order to remove the contamination from the water. So far known this has reduced PFOA concentrations in water considerably, however, no measurements and/or levels were available at time of publication.

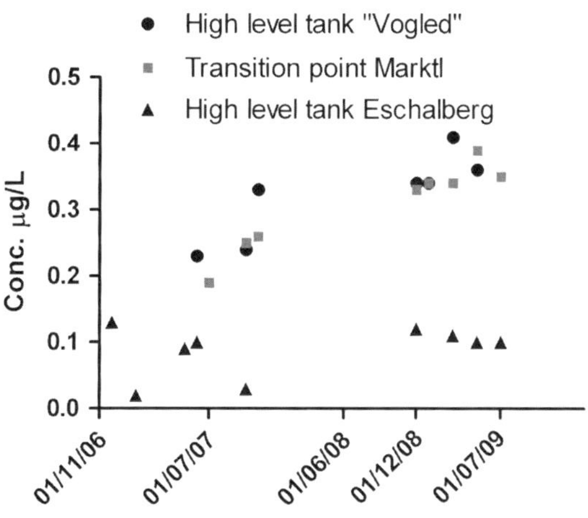

Fig. 9 Concentration of PFOA in Bavarian (Germany) drinking water from the Inn-Salzach group (high level tank "Vogled" and transition point Marktl) and the communities of Burgkirchen and Emmerting (high level tank Eschalberg)

A monitoring study on the presence of PFOA and PFOS in tap water from 20 sites across England was carried out in the course of 2007 [67]. PFOS was found at four sites at relatively constant levels over time. The highest levels of PFOS (162 ng/L) were observed south of Cambridge in groundwater near an airstrip, confirming that airstrips are a potential source of PFCs to the environment. One other sampled site was near the Buncefield site where a series of large explosions followed by a big fire occurred in oil storage tanks in December 2005 and was fought with large volumes of AFFFs. It appeared that the groundwater pumping station in the vicinity of the explosion site providing raw water for drinking water supply was contaminated with PFCs. Although activated carbon treatment was included in the water treatment, concentrations in the effluent from the station amounted to 66 ng/L for PFOA and 45 ng/L for PFOS in one occasion. According to Atkinson et al. [67], the activated carbon beds were not regenerated for some years, which seem to be a reasonable explanation for the relatively high levels of PFCs encountered in the drinking water. Temporal and spatial variations across the sites were relatively high. Minimum and maximum concentrations measured were between 25 and 370 ng/L of PFOA, meaning that the guidance value (tier 1) of 300 ng/L for PFOA levels in drinking water set by the DWI [88] was exceeded on one occasion, which according to [88] triggers further monitoring and consultation with local health authorities.

Another case of drinking water contamination by PFOS was found in the vicinity of an airstrip in East Anglia in England. PFHxA, PFOA, PFHxS and PFOS concentrations in the source water varied around 500, 1,000, 1,500 and 2,500 ng/L over a measuring period of 2 years. In order to remove PFCs from the raw water activated carbon filters were installed. To increase removal efficiencies, water/GAC contact times were increased from 30 min to between 65 and 110 min and regeneration frequency was increased from biennial to annual (5,500 bed volumes between regeneration). PFOS was readily removed from the raw water and effluent PFOS concentrations were generally below the LOQ of 100 ng/L [89]. PFHxA was the first compound to break through after approximately 2,000 bed volumes and was the least readily removed compound. PFOA, PFHxS, and PFOS showed breakthrough after more than 5,500 bed volumes.

A similar behavior of PFCs was well documented in a drinking water treatment plant in Oakdale, USA. The tap water produced in this plant, the influent water of which is contaminated with PFCs (see Chap. 3 and Fig. 10), has been monitored extensively over the past few years. From Fig. 10 it can be concluded that the short chained PFBA, PFPeA, and PFHxA are not well retained by the treatment plant. This can be seen at early 2007 and early 2009 when the PFCs break through the GAC filter. By the end of 2008, the GAC was regenerated and fresh GAC retained PFCs well for a short period of time. Other PFCs (PFBS, PFHxS and PFOS) were not detected in the treatment plant effluent drinking water.

After several pollution incidences became known, guideline values have been set in the recent past by the Drinking Water Inspectorate (DWI) of England and Wales, the German Drinking Water Commission and by authorities in the USA. A review of these values was given by Rumsby et al. [72]. However, guideline

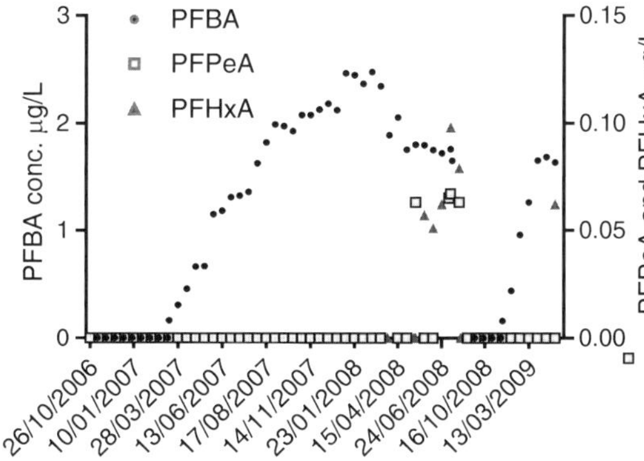

Fig. 10 PFBA, PFPeA, and PFHxA concentrations in the combined GAC effluent of a drinking water production plant in Oakdale, USA (courtesy of the Minnesota Department of Health, personal communication with Kolstad Chad)

values vary between countries. For example, for a lifelong exposure the combined PFOS and PFOA concentrations of 300 ng/L should not be exceeded in Germany [87] whereas individual values of 300 ng/L for PFOA and 300 ng/L for PFOS are used as the lowest guidance levels of a three-tiered system in the UK, where minimum action has to be taken by monitoring and consultation with the local health professionals [88]. Recently, provisional health related indication values (HRIV) were also proposed for short chain PFCs, e.g., 3,000 ng/L for PFBS and 7,000 ng/L for PFBA [71].

4.2 *Behavior of PFCs During Drinking Water Preparation*

In order to reduce the PFC concentrations of contaminated raw waters below the recommended health based values different options exist. The removal efficiency of the different PFCs from water during treatment is strongly dependent on the type of treatment processes used and on the chain length and nature of head groups of the PFCs. Depending on the applied treatment, it was found that PFCs may be present up to the same level in the drinking water as in the source water. This finding demonstrates that PFC removal efficiencies in the drinking water treatment process in general are low [82, 90]. Different studies showed that there are not only problems with groundwater sources, but also a correlation between surface water and tap water from the same region [33, 91]. Natural processes like river bank filtration or dune filtration are ineffective [92]. Lange et al. [93] studied the concentrations of PFCs in the Rhine and compared them to concentrations after river bank filtration. Typical concentrations were in the low ng/L range and riverbank filtration did not remove the PFCs. This has recently been confirmed by a survey of influent and effluent concentrations of several drinking water treatment plants in the USA [90] and in pre-treated infiltrated Rhine water in dune areas used as a treatment step in the drinking water production where water had travel times up to 18 years [92].

As described in part above, at present, technical measures taken in order to remove the PFCs from the raw water are almost invariably the use of GAC filters. The order of breakthrough of PFCs is increasing with decreasing chain length and appears to be faster for carboxylates than sulfonates.

In a recent study, which analyzed influent and effluent concentrations from drinking water treatment plants, it was concluded that only the treatment plants with membrane filtration removed PFCs efficiently [90]. However, PFCs analyzed did not include compounds with carbon chain lengths shorter than C6, thus not revealing the removal capacity for, e.g., PFBA and PFBS at process scale [90]. The generation of a concentrated waste stream when membrane filtration is used and the relatively high operation costs make this treatment method not widely used yet in the drinking water treatment process.

The description in the literature of the different processes and sorption parameters still is vague and sometimes contradicting. However, it appears that the regeneration rate of GAC columns and the contact time of the water with the activated carbon are important parameters in the efficient removal of PFCs from water. For further detailed information on water treatment options for PFC removal see Chap. "Treatment options for the Removal and Degradation of Polyfluorinated Chemicals" of this book.

5 Summary

The presence of PFCs at a base level of contamination due to pollution from diffuse sources and global/continental distribution may occur nowadays. The background level in many European rivers has been known for some years. The source of PFCs in the environment can usually be traced to a discharging factory, accidental spill or wastewater treatment plant.

PFC concentrations in the Central and Southern European rivers, such as in Italy, Germany, The Netherlands, and UK, generally seem to be higher than in Northern Europe. This is well illustrated when levels reported for the Scandinavian countries and Northern Poland. However, this conclusion might also be an artifact due to the situation that in some countries more analyses were carried out than in others, and thus the possibility of hot spot identification is higher. The rivers Po, Rhine and Seine appear to be the major rivers in Europe discharging PFCs into the oceans [3]. The reports generally focus on the presence of PFOA and PFOS in the environment. However, as a result of substitution of C8 compounds by C4 perfluorinated and polyfluorinated telomer compounds, respectively, it is expected that concentrations of the substitutes or their metabolites will increase in the environment. Unfortunately, PFBA and PFBS have been monitored only scarcely thus far.

Concentrations in drinking waters remain on average fairly low. Drinking water produced from raw water extracted in the vicinity of a PFC spill tends to be contaminated. As for the removal of PFCs during drinking water preparation several conclusions can be drawn. In practice, two technologies known to remove

PFCs also used in the drinking water treatment process are membrane and activated carbon filtration. The difference in PFC baseline concentrations in drinking water will depend on the technologies used in different treatment plants. Drinking water prepared by a treatment which does not include GAC filtration or reverse osmosis will generally contain higher PFCs levels in the case contaminated water is used as source water.

Acknowledgment We thank Dr. Beatrix Vieten for her kind assistance in the preparation of the map in Figure 1.

References

1. Hekster FM, Laane RWPM, De Voogt P (2003) Environmental and toxicity effects of perfluoroalkylated substances. Rev Environ Contamin Toxicol 179:99–121
2. Skutlarek D, Exner M, Farber H (2006) Perfluorinated surfactants in surface and drinking waters. Environ Sci Pollut 13(5):299–307
3. McLachlan MS, Holmstrom KE, Reth M, Berger U (2007) Riverine discharge of perfluorinated carboxylates from the European continent. Environ Sci Technol 41(21): 7260–7265
4. Moody CA, Martin JW, Kwan WC, Muir DCG, Mabury SC (2002) Monitoring perfluorinated surfactants in biota and surface water samples following an accidental release of fire-fighting foam into Etobicoke Creek. Environ Sci Technol 36(4):545–551
5. Moody CA, Hebert GN, Strauss SH, Field JA (2003) Occurrence and persistence of perfluorooctanesulfonate and other perfluorinated surfactants in groundwater at a fire-training area at Wurtsmith Air Force Base, Michigan, USA. J Environ Monit 5(2):341–345
6. Murakami M, Kuroda K, Sato N, Fukushi T, Takizawa S, Takada H (2009) Groundwater pollution by perfluorinated surfactants in Tokyo. Environ Sci Technol 43(10):3480–3486
7. Ericson I, Domingo JL, Nadal M, Bigas E, Llebaria X, van Bavel B, Lindstrom G (2009) Levels of perfluorinated chemicals in municipal drinking water from catalonia Spain: public health implications. Arch Environ Contam Toxicol 57(4):631–638
8. Ericson I, Nadal M, Van Bavel B, Lindstrom G, Domingo JL (2008) Levels of perfluorochemicals in water samples from Catalonia, Spain: Is drinking water a significant contribution to human exposure? Environ Sci Pollut Res 15(7):614–619
9. Ahrens L, Barber JL, Xie ZY, Ebinghaus R (2009) Longitudinal and latitudinal distribution of perfluoroalkyl compounds in the surface water of the Atlantic Ocean. Environ Sci Technol 43(9):3122–3127
10. Yamashita N, Kannan K, Taniyasu S, Horii Y, Petrick G, Gamo T (2005) A global survey of perfluorinated acids in oceans. Mar Pollut Bull 51:658–668
11. Higgins CP, Luthy RG (2006) Sorption of perfluorinated surfactants on sediments. Environ Sci Technol 40(23):7251–7256
12. Becker AM, Gerstmann S, Frank H (2008) Perfluorooctanoic acid and perfluorooctane sulfonate in the sediment of the Roter Main river, Bayreuth, Germany. Environ Pollut 156 (3):818–820
13. Furdui VI, Helm PA, Crozier PW, Lucaciu C, Reiner EJ, Marvin CH, Whittle DM, Mabury SA, Tomy GT (2008) Temporal trends of perfluoroalkyl compounds with isomer analysis in lake trout from Lake Ontario (1979–2004). Environ Sci Technol 42(13):4739–4744
14. Butt CM, Mabury SA, Kwan M, Wang XW, Muir DCG (2008) Spatial trends of perfluoroalkyl compounds in ringed seals (Phoca hispida) from the Canadian Arctic. Environ Toxicol Chem 27(3):542–553

15. Dai JY, Li M, Jin YH, Saito N, Xu MQ, Wei FW (2006) Perfluorooctanesulfonate and perfluorooctanoate in red panda and giant panda from China. Environ Sci Technol 40(18): 5647–5652
16. Vestergren R, Cousins IT (2009) Tracking the pathways of human exposure to perfluorocarboxylates. Environ Sci Technol 43(15):5565–5575
17. Kannan K, Corsolini S, Falandysz J, Fillmann G, Kumar KS, Loganathan BG, Mohd MA, Olivero J, Van Wouwe N, Yang JH, Aldous KM (2004) Perfluorooctanesulfonate and related fluorochemicals in human blood from several countries. Environ Sci Technol 38 (17):4489–4495
18. Kissa E (2001) Fluorinated surfactants and repellents, vol 97, 2nd edn. Marcel Dekker, New York, p 615
19. U.S. Environmental Protection Agency: 2010/2015 PFOA Stewardship Program. http://www.epa.gov/oppt/pfoa/pubs/stewardship/ (02-02-2011)
20. Stockholm convention on POPs. http://chm.pops.int/Programmes/NewPOPs/The9newPOPs/tabid/672/language/en-US/Default.aspx (02-02-2011)
21. EU, Directive 2006/122/EC of the European Parliament and of the Council of 12 December 2006. Official Journal of the European Union 2006
22. Olsen GW, Mair DC, Church TR, Ellefson ME, Reagen WK, Boyd TM, Herron RM, Medhdizadehkashi Z, Nobilett JB, Rios JA, Butenhoff JL, Zobel LR (2008) Decline in perfluorooctanesulfonate and other polyfluoroalkyl chemicals in American Red Cross adult blood donors, 2000–2006. Environ Sci Technol 42(13):4989–4995
23. Prevedouros K, Cousins IT, Buck RC, Korzeniowski SH (2006) Sources, fate and transport of perfluorocarboxylates. Environ Sci Technol 40(1):32–44
24. Schultz MM, Barofsky DF, Field JA (2004) Quantitative determination of fluorotelomer sulfonates in groundwater by LC MS/MS. Environ Sci Technol 38(6):1828–1835
25. Sinclair E, Kannan K (2006) Mass loading and fate of perfluoroalkyl surfactants in wastewater treatment plants. Environ Sci Technol 40(5):1408–1414
26. Boulanger B, Vargo JD, Schnoor JL, Hornbuckle KC (2005) Evaluation of perfluorooctane surfactants in a wastewater treatment system and in a commercial surface protection product. Environ Sci Technol 39(15):5524–5530
27. Bossi R, Strand J, Sortkjaer O, Larsen MM (2008) Perfluoroalkyl compounds in Danish wastewater treatment plants and aquatic environments. Environ Int 34(4):443–450
28. Eggen T, Moeder M, Arukwe A (2010) Municipal landfill leachates: a significant source for new and emerging pollutants. Sci Total Environ 408(21):5147–5157
29. Busch J, Ahrens L, Sturm R, Ebinghaus R (2010) Polyfluoroalkyl compounds in landfill leachates. Environ Pollut 158(5):1467–1471
30. Ellis DA, Martin JW, De Silva AO, Mabury SA, Hurley MD, Andersen MPS, Wallington TJ (2004) Degradation of fluorotelomer alcohols: a likely atmospheric source of perfluorinated carboxylic acids. Environ Sci Technol 38(12):3316–3321
31. Loewen M, Wania F, Wang FY, Tomy G (2008) Altitudinal transect of atmospheric and aqueous fluorinated organic compounds in Western Canada. Environ Sci Technol 42 (7):2374–2379
32. Murakami M, Shinohara H, Takada H (2009) Evaluation of wastewater and street runoff as sources of perfluorinated surfactants (PFSs). Chemosphere 74(4):487–493
33. Lien NPH, Fujii S, Tanaka S, Nozoe M, Wirojanagud W, Anton A, Lindstrom G (2006) Perfluorinated substances in tapwater of Japan and several countries and their relationship to surface water contamination. Environ Eng Res 43:611–618
34. Kallenborn R, Berger U, Jarnberg U (2004) Perfluorinated alkylated substances (PFAS) in the Nordic environment. Nordic Council of Ministers, Copenhagen, Denmark
35. Ahrens L, Felizeter S, Sturm R, Xie Z, Ebinghaus R (2009) Polyfluorinated compounds in waste water treatment plant effluents and surface waters along the River Elbe, Germany. Mar Pollut Bull 58(9):1326–1333

36. Ahrens L, Plassmann M, Xie ZY, Ebinghaus R (2009) Determination of polyfluoroalkyl compounds in water and suspended particulate matter in the river Elbe and North Sea, Germany. Front Environ Sci Eng China 3(2):152–170
37. Rostkowski P, Taniyasu S, Yamashita N, Falanddysz J, Zegarowsky L, Chojnacka A, Pazdro K, Falanddysz J (2009) Survey of perfluorinated compounds (PFCs) in surface waters of Poland. J Environ Sci Health Part A 44:1518–1527
38. http://www.gendorf.de/ (22.11.2007)
39. Wasserwirtschaftsamt Traunstein: Hydrogeologisches Basisgutachten Grundwasserströmungsverhältnisse im Öttinger Forst; Bayerisches Landesamt für Umwelt (2009)
40. Wilhelm M, Bergmann S, Dieter HH (2010) Occurrence of perfluorinated compounds (PFCs) in drinking water of North Rhine-Westphalia, Germany and new approach to assess drinking water contamination by shorter-chained C4-C7 PFCs. Int J Hyg Environ Health 213 (3):224–232
41. http://www.lfu.bayern.de/analytik_stoffe/analytik_org_stoffe_perfluorierte_chemikalien/pfc_belastung_oberflaechengewaesser_talsperren/index.htm (03-08-11)
42. Möller M, Ahrens L, Sturm R, Ebinghaus R (2009) Identification of point sources of polyfluoroalkyl compounds (PFCs) along the River Rhine watershed and their transportation into the North Sea. Coastline Rep 13:143–154
43. ARW, Jahresbericht 2008. 2009; 65
44. http://www.nrw.de/presse/pft-im-rheder-bach-gefunden-2119/ (02-02-2011)
45. http://www.nrw.de/presse/verursacher-der-pft-belastung-in-rhede-gefunden-2176/ (02-02-2011)
46. RIWA Jaarrapport 2009: De Rijn; RIWA Rijnwaterbedrijven 2010
47. RIWA Jaarrapport 2008: De Rijn; RIWA Rijnwaterbedrijven: 2009
48. de Voogt P, Berger U, de Coen W, de Wolf W, Heimstad E, MCLachlan MS, van Leeuwen S, van Roon A (2006) Perfluorinated organic compounds in the European environment; Institute for biodiversity and ecosystem dynamics, Universiteit van Amsterdam
49. RIWA Jaarrapport 2007: De Rijn; RIWA Rijnwaterbedrijven 2008
50. RIWA Jaarrapport 2006: De Rijn; RIWA Rijnwaterbedrijven 2007
51. Kwadijk CJAF, Korytár P, Koelmans AA (2010) Distribution of perfluorinated compounds in aquatic systems in the Netherlands. Environ Sci Technol 44(10):3746–3751
52. Loos R, Gawlik BM, Locoro G, Rimaviciute E, Contini S, Bidoglio G (2009) EU-wide survey of polar organic persistent pollutants in European river waters. Environ Pollut 157 (2):561–568
53. Ferrey M, Wilson JT, Adair C (2009) Extent of sorption and biodegradability of perfluorooctanoic acid (PFOA) and perfluorooctanesulfonic acid (PFOS) in aquifer sediment, In Situ and On-Site Bioremediation-2009: Proceedings of the 10th International In Situ and On-Site Bioremediation Symposium, 2009
54. Cheng J, Vecitis CD, Park H, Mader BT, Hoffmann MR (2010) Sonochemical degradation of perfluorooctane sulfonate (PFOS) and perfluorooctanoate (PFOA) in groundwater: kinetic effects of matrix inorganics. Environ Sci Technol 44(1):445–450
55. Bartell SM, Calafat AM, Lyu C, Kato K, Ryan PB, Steenland K (2010) Rate of decline in serum PFOA concentrations after granular activated carbon filtration at two public water systems in Ohio and West Virginia. Environ Health Perspect 118(2):222–228
56. Loganathan BG, Sajwan KS, Sinclair E, Kumar KS, Kannan K (2007) Perfluoroalkyl sulfonates and perfluorocarboxylates in two wastewater treatment facilities in Kentucky and Georgia. Water Res 41(20):4611–4620
57. Pistocchi A, Loos R (2009) A map of European emissions and concentrations of PFOS and PFOA. Environ Sci Technol 43(24):9237–9244
58. Lange FT, Schmidt CK, Brauch HJ (2007) Perfluorinated surfactants: the perfluorooctanesulfonate (PFOS) substitute perfluorobutanesulfonate (PFBS) increasingly affects the raw water quality of Rhine waterworks. GWF Wasser Abwasser 148(7–8):510–516
59. RIWA *Jaarrapport 2008: De Rijn*; RIWA Rijnwaterbedrijven 2008

60. Möller M, Ahrens L, Surm R, Westerveld J, van der Wielen F, Ebinghaus R, de Voogt P (2010) Distribution and sources of polyfluoroalkyl substances (PFAS) in the River Rhine watershed. Environ Pollut 158:3243–3250
61. DVGW research project W 7/01/04. Investigations on the occurrence of perfluorinated alkyl compounds in German drinking water resources, final report; 2007; p 82
62. Van Leeuwen (2009), S. Bluswater calamiteit Schiphol juli 2008: Monitoring van PFOS in het oppervlaktewater; Instituut voor Milieuvraagstukken
63. Hangen E, Klemm A, Kronawitter H, Schubert A (2010) Perfluorooctanoate (PFO) in forest soils near a fluoropolymer manufacturing facility. Water Air Soil Pollut 212(1):491–499
64. Loos R, Locoro G, Huber T, Wollgast J, Christoph EH, De Jager A, Gawlik BM, Hanke G, Umlauf G, Zaldivar JM (2008) Analysis of perfluorooctanoate (PFOA) and other perfluorinated compounds (PFCs) in the River Po watershed in N-Italy. Chemosphere 71 (2):306–313
65. Valsecchi, S.; Polesello, S., The search of sources of perfluoroalkyl acids (PFAAs) in Northern Italian waters. In *SETAC 19th annual meeting*, Goteborg, 2009; Poster presentation; abstract book, p. 129
66. http://www.environment-agency.gov.uk/homeandleisure/pollution/water/89141.aspx (02-02-2011)
67. Atkinson C, Blake S, Hall T, Kanda K, Rumsby P (2008) Survey of the prevalence of perfluorooctane sulphonate (PFOS), Perfluorooctanoic acid (PFOA) and related compounds in drinking water and their sources; DEFRA 7585; Drinking Water Inspectorate, Department for Environment, Food and Rural Affairs
68. Schultz MM, Barofsky DF, Field JA (2003) Fluorinated alkyl surfactants. Environ Eng Sci 20(5):487–501
69. Moody CA, Field JA (2000) Perfluorinated surfactants and the environmental implications of their use in fire-fighting foams. Environ Sci Technol 34(18):3864–3870
70. Levine AD, Libelo EL, Bugna G, Shelley T, Mayfield H, Stauffer TB (1997) Biogeochemical assessment of natural attenuation of JP-4-contaminated ground water in the presence of fluorinated surfactants. Sci Total Environ 208(3):179–195
71. Wilhelm M, Bergmann S, Dieter HH (2010) Occurrence of perfluorinated compounds (PFCs) in drinking water of North Rhine-Westphalia, Germany and new approach to assess drinking water contamination by shorter-chained C4-C7 PFCs. Int J Hyg Environ Health 213(3):224–232
72. Rumsby PC, McLaughlin CL, Hall T (2009) Perfluorooctane sulphonate and perfluorooctanoic acid in drinking and environmental waters. Philos Trans Royal Soc Math Phys Eng Sci (367):4119–4136
73. *Landesanstalt für Umwelt, Messungen und Naturschutz Baden-Württemberg (LUBW): Grundwasser-Überwachungsprogramm, Ergebnisse der Beprobung 2006*. 2007; Vol./Bd. 34. See: http://www.lubw.baden-wuerttemberg.de/servlet/is/30330/grundwasser_ueberwachung_ergebnisse_2006.pdf?command=downloadContent&filename=grundwasser_ueberwachung_ergebnisse_2006.pdf (03-08-11)
74. Mons MN, de Voogt P, van Roon A (2007) Perfluoroalkylated substances in Dutch drinking water sources; BTO 2007.048.; KWR: Nieuwegein
75. http://www.perfood.org (02-02-11)
76. Tardiff RG, Carson ML, Sweeney LM, Kirman CR, Tan YM, Andersen M, Bevan C, Gargas ML (2009) Derivation of a drinking water equivalent level (DWEL) related to the maximum contaminant level goal for perfluorooctanoic acid (PFOA), a persistent water soluble compound. Food Chem Toxicol 47(10):2557–2589
77. Holzer J, Goen T, Rauchfuss K, Kraft M, Angerer J, Kleeschulte P, Wilhelm M (2009) One-year follow-up of perfluorinated compounds in plasma of German residents from Arnsberg formerly exposed to PFOA-contaminated drinking water. Int J Hyg Environ Health 212(5): 499–504

78. Wilhelm M, Holzer J, Dobler L, Rauchfuss K, Midasch O, Kraft M, Angerer J, Wiesmuller G (2009) Preliminary observations on perfluorinated compounds in plasma samples (1977–2004) of young German adults from an area with perfluorooctanoate-contaminated drinking water. Int J Hyg Environ Health 212(2):142–145
79. Emmett EA, Zhang H, Shofer FS, Freeman D, Rodway NV, Desai C, Shaw LM (2006) Community exposure to perfluorooctanoate: Relationships between serum levels and certain health parameters. J Occ Environ Med 48(8):771–779
80. EFSA (2008) Scientific Opinion of the Panel on Contaminants in the Food Chain Perfluorooctane sulfonate (PFOS), perfluorooctanoic acid (PFOA) and their salts. EFSA J 653:1–131
81. Lange FT, Brauch HJ (2008) Perfluorierte Verbindungen in deutschen Rohwasserressourcen. Energie Wasser-Praxis
82. Loos R, Wollgast J, Huber T, Hanke G (2007) Polar herbicides, pharmaceutical products, perfluorooctanesulfonate (PFOS), perfluorooctanoate (PFOA), and nonylphenol and its carboxylates and ethoxylates in surface and tap waters around Lake Maggiore in Northern Italy. Anal Bioanal Chem 387(4):1469–1478
83. D'Hollander W, Roosens L., Covaci A, Ornelis C, Smolders R, Van den Heuvel R, de Voogt P, Nobels I, Van Parys C, De Coen W, Bervoets L (2009) Gebromeerde brandvertragers en perfluorverbindingen in Vlaanderen: onderzoek naar verspreiding, humane opname, gehaltes in humane weefsels en/of lichaamsvochten, en gezondheidseffecten als basis voor de selectie van geschikte milieu en gezondheidsindicatoren (BFRISK); Departement van Leefmilieu, natuur en energie
84. Haug LS, Salihovic S, Jogsten IE, Thomsen C, van Bavel B, Lindstrom G, Becher G (2010) Levels in food and beverages and daily intake of perfluorinated compounds in Norway. Chemosphere 80(10):1137–1143
85. http://www.pft.lua.nrw.de/owl/GIS/exhibit/pft_tw.php?exhibit-use-local-resources#2 (02-02-2011)
86. http://www.lgl.bayern.de/gesundheit/umweltmedizin/perfluorierte_tenside_altoetting.htm (02-02-2011)
87. TWK (Drinking Water Commission of the German Ministry of Health at the German Federal Environment Agency, 2006). Provisional evaluation of PFT in drinking water with the guide substances perfluorooctanoicacid (PFOA) and perfluorooctane sulfonate (PFOS) as examples. See: http://www.umweltbundesamt.de/uba-info-presse-e/hintergrund/pft-in-drinking-water.pdf (02-02-2011)
88. DWI (Drinking Water Inspectorate). Guidance on the Water Supply (Water Quality) Regulations 2000 specific to PFOS (perfluorooctane sulphonate) and PFOA (perfluorooctanoic acid) concentrations in drinking water; 2009
89. Markall B (2008) Removal of perfluorinated compounds by GAC treatment. Presentation at Second European Water and Wastewater Management Conference, Birmingham, UK
90. Quinones O, Snyder SA (2009) Occurrence of perfluoroalkyl carboxylates and sulfonates in drinking water utilities and related waters from the United States. Environ Sci Technol 43(24):9089–9095
91. Takagi S, Adachi F, Miyano K, Koizumi Y, Tanaka H, Mimura M, Watanabe I, Tanabe S, Kannan K (2008) Perfluorooctanesulfonate and perfluorooctanoate in raw and treated tap water from Osaka Japan. Chemosphere 72(10):1409–1412
92. Eschauzier C, Haftka J, Stuyfzand PJ, De Voogt P (2010) Perfluorinated compounds in infiltrated river Rhine water and infiltrated rainwater in coastal dunes. Environ Sci Technol 44(19):7450–7455
93. Lange FT, Wenz M, Schmidt CK, Brauch HJ (2007) Occurrence of perfluoroalkyl sulfonates and carboxylates in German drinking water sources compared to other countries. Water Sci Technol 56(11):151–158
94. AWBR, Jahresbericht 2007. 2008;39
95. Weremiuk, AM, Gerstmann, S, Frank, H (2006) Quantitative determination of perfluorinated surfactants in water by LC-ESI-MS/MS. J Sep Sci 29(14):2251–2255

96. AWBR, Jahresbericht 2008. 2009;40
97. AWBR, Jahresbericht 2009. 2010;41
98. ARW, Jahresbericht 2007. 2008;64
99. Clara M, Gans O, Weiss S, Sanz-Escribano D, Scharf S, Scheffknecht C (2009) Perfluorinated alkylated substances in the aquatic environment: an Austrian case study. Water Res 43(18):4760–4768
100. Huset CA, Chiaia AC, Barofsky DF, Jonkers N, Kohler HPE, Ort C, Giger W, Field JA (2008) Occurrence and mass flows of fluorochemicals in the Glatt Valley watershed, Switzerland. Environ Sci Technol 42(17):6369–6377

Treatment Options for the Removal and Degradation of Polyfluorinated Chemicals

Holger Lutze, Stefan Panglisch, Axel Bergmann, and Torsten C. Schmidt

Abstract This chapter deals with different treatment options for the removal or degradation of polyfluorinated chemicals (PFC). Adsorption on activated carbon and membrane filtration (nanofiltration and reverse osmosis) belongs to the state-of-the-art methods and effective separate resp. reject fluorinated compounds. Biological degradation and conventional oxidative techniques for pollutant control such as advanced oxidation (ozonation, UV/H_2O_2, Fenton process) seem not to be suitable for PFC degradation. New approaches for the oxidation of fluorinated chemicals are based on the formation of sulfate radical anions (e.g., by photolysis of peroxodisulfate), sonolysis, and electrolysis with boron-doped diamond electrodes. Some approaches regarding reductive treatment have been reported to degrade PFC. However, hardly any information about by-product formation and degradation efficiency under real conditions are available regarding these new oxidation and reduction techniques.

Keywords Degradation • Polyfluorinated compounds • Sorption • Water treatment

Contents

H. Lutze (✉) • T.C. Schmidt
Instrumental Analytical Chemistry, University Duisburg-Essen, Universitätsstr. 5, 45141 Essen, Germany

IWW Water Centre, Moritzstr. 26, 45476 Mülheim a.d. Ruhr, Germany
e-mail: holger.lutze@uni-due.de

S. Panglisch • A. Bergmann
IWW Water Centre, Moritzstr. 26, 45476 Mülheim a.d. Ruhr, Germany

T.P. Knepper and F.T. Lange (eds.), *Polyfluorinated Chemicals and Transformation Products*, Hdb Env Chem (2012) 17: 103–125
DOI 10.1007/978-3-642-21872-9_6,

Abbreviations

$^{\bullet}OH$	Hydroxyl radical
AC	Activated carbon
AOP	Advanced oxidation processes
BDDE	Boron-doped diamond electrodes
CMC	Critical micelle concentration
DOC	Dissolved organic matter
Fe(0)	Elemental iron
GAC	Granular activated carbon
H4PFOS	*1H,1H,2H,2H*-perfluorooctane sulfonate
HSO_5^-	Peroxomonosulfate
K_C	Freundlich constant for carbon mass of the molecule
K_F	Freundlich constant
K_L	Langmuir constant
LP Hg lamp	Low pressure mercury lamp
M_C	Carbon mass of a molecule
M_F	Microfiltration
n	Freundlich exponent
N-EtFOSAA	2(*N*-ethyl-perfluorooctanesulfonamido) acetic acid
N-EtFOSE	2(*N*-ethyl-perfluorooctanesulfonamido) ethyl alcohol
NF	Nanofiltration
PAC	Powdered activated carbon
PAC-0.8	Powdered activated carbon with a mean particle size of 0.8 μm
PAC-10	Powdered activated carbon with a mean particle size of 10 μm
PFA	Perfluorocarboxylic acid
PFBA	Perfluorobutanoic acid
PFBS	Perfluorobutane sulfonic acid
PFC	Per- and polyfluorinated compounds
PFHxA	Perfluorohexanoic acid
PFOA	Perfluorooctanoic acid

PFOS	Perfluorooctane sulfonic acid
PFS	Perfluorosulfonic acid
q	Equilibrium load in Langmuir model
qm	Maximal load in Langmuir model
RO	Reverse osmosis
$S_2O_8^{2-}$	Peroxodisulfate
$SO_4^{\bullet -}$	Sulfate radical anion
UF	Ultrafiltration
VUV	Vacuum UV (wavelength <200 nm)
Xe–Hg-Lamp	Xenon-doped mercury lamp

1 Introduction

The distribution, environmental behavior, human health risk, and emission routes of Per- and polyfluorinated compounds (PFC) are intensively discussed in science as well as on a political level. Due to the potential harm for human health, the US-EPA proposed an advisory drinking water standard for perfluorooctanoic acid (PFOA) and perfluorooctane sulfonic acid (PFOS) of 0.4 $\mu g\ l^{-1}$ [1]. An advisory threshold value for the sum of PFOA and PFOS has also been set by the German drinking water commission (0.3 $\mu g\ l^{-1}$) [2]. The ongoing discussion about micropollutant control suggests that a regulation for PFC in domestic wastewater treatment will be set in the near future. Very little is known about the consequences of elevating perfluorochemical concentrations in the environment and how this is connected with health and economical risks. This is aggravated by the high persistence of these compounds in the environment and the tendency for bioaccumulation especially of long-chain PFC like PFOA and PFOS [3]. Due to the high mobility and ubiquitous occurrence of PFC, the remediation of contaminated sites is very difficult. Thus the prevention of PFC release is important that is partly achieved by the development and usage of alternative agents. However, the unique features of perfluorinated organics are still important for some applications such as fire-fighting foams [3] and the production of semiconductors [4] and thus renders replacement difficult. In addition to drinking water, other routes of exposure have also to be taken into account. Drinking water is probably not the main source for PFC exposure to humans, whereas domestic dust, food and textiles probably play a more important role within this context [5].

Perfluoro chemicals are present in the environment and have been detected all over the globe [6–14], whereby fluorotelomer alcohols may act as highly mobile precursors for perfluorinated carboxylic acids [8]. PFC are present in drinking water resources, where they probably persist for a long time due to their high environmental stability. Thus drinking water suppliers have to deal with the possibility of elevated PFC concentrations in their raw water and thereby need to consider treatment strategies as barriers for PFC.

PFC survive most of the conventional techniques in drinking and wastewater treatment and are found in finished drinking water in Germany, Switzerland, USA, and other countries [13, 15]. In some of these cases, the PFC concentrations exceeded the advisory drinking water standard suggested by the German and US-EPA drinking water commission.

In particular, treatment techniques based on the structural change of the target molecules such as ozonation or advanced oxidation fail due to the high chemical stability of these compounds [16]. Separation methods such as ion exchange and sorption on activated carbon as well as nanofiltration and reverse osmosis appear to be effective to remove PFC from water. The waste produced is enriched with the pollutants and needs to be treated further which can be done via incineration. The following chapter reviews current strategies for the treatment of water contaminated with fluorinated chemicals and indicates new trends in this sector.

2 Physical Treatment

2.1 *Adsorption*

Adsorption on activated carbon and ion exchange belongs to the state-of-the-art techniques to treat water containing PFC. The sorption of these compounds on activated carbon and ion exchange resins can be described with Langmuir and Freundlich isotherms. The corresponding constants found in the literature are summarized in Table 1.

Table 1 Constants for Langmuir and Freundlich isotherms for different perfluorinated compounds and adsorbents

PFC	Sorbent name	Type of sorbent	K_L (l mg^{-1})	qm (mg g^{-1})	K_F ((mg/g) $(mg/l)^{-1}$)	K_C ((mgC/g) $(mgC/l)^{-1}$)	*n*
PFOS[17]	GAC Filtrasorb F300	GAC	0.068	196.2	39	13[a]	0.33
PFOS[17]	URV Mod1	PAC	0.08	211.6	37	13[a]	0.37
PFOS[17]	Filtrasorb F400	PAC	0.124	236.4	61	19[a]	0.29
PFOA[17]	GAC Filtrasorb F400	GAC	0.038	112.1	12	5[a]	0.44
PFBS[17]	GAC Filtrasorb F400	GAC	0.034	98.7	9	5[a]	0.46
PFOS[18]	–	GAC	–	–	56[a]	14[a]	0.18
PFOS[18]	–	PAC	–	–	165[a]	43[a]	0.18
PFOS[18]	–	Anion exchange resin[b]	–	–	169[a]	43[a]	0.17
PFOA[18]	–	GAC	–	–	29[a]	10[a]	0.28
PFOA[18]	–	PAC	–	–	123[a]	38[a]	0.20
PFOA[18]	–	Anion exchange resin[b]	–	–	636[a]	178[a]	0.13

[a]Calculated from data available in literature, [b]Strongly basic gel-type resin, quaternary ammonium functionality (Amberlite IRA 400 resin), K_L Langmuir constant, *qm* maximal load, K_F Freundlich constant regarding the molecular mass, K_C Freundlich constant regarding the mass of carbon, *n* Freundlich exponent, *PFOA* perfluorooctanoic acid, *PFOS* perfluorooctane sulfonic acid, *PFBS* perfluorobutane sulfonic acid, *GAC* granular activated carbon, *PAC* powdered activated carbon

A criterion often used to characterize the sorptivity of a pollutant on a sorbent is the Freundlich constant (K_F), which can be normalized to the carbon mass (M_C) of the target molecule (M_C (PFOA and PFOS): 96 g carbon mol^{-1}). Based on the sorption affinity of different fractions of natural organic matter on activated carbon, K_C values of below 20 (mg C/g) (mg C/l)$^{-1}$ indicate a poor adsorbability, whereas compounds with $K_C > 50$ (mg C/g) (mg C/l)$^{-1}$ can be considered as strongly adsorbable (derived from [19]).

The K_C values of < 20 (mg C/g) (mg C/l)$^{-1}$ for PFOA, PFOS, and PFBS indicate a low sorptivity for all three compounds on GAC. Thus, the efficiency of the sorption process is particularly sensitive toward competitive sorption of DOC (e.g., Ruhr River, Mülheim (Germany), pH 7.6, 54% of DOC strongly adsorbable: K_C 55 (mgC/g) (mgC/l)$^{-1}$, 31.8% of DOC poorly adsorbable: K_C 17 (mgC/g) (mgC/l)$^{-1}$, 14.1% of DOC nonadsorbable: $K_C < 17$ (mgC/g) (mgC/l)$^{-1}$ [19]). The data shown in [18] indicate that powdered activated carbon could reveal much higher loads under equilibrium conditions as well as faster sorption kinetics for PFOS and PFOA. This is probably due to a more efficient transport into the inner micropore system, which provides most of the sorption sites [19]. However, the adsorbability depends also on the type of carbon used [19], thus activated carbons with higher affinities toward these compounds may exist.

GAC is thermally reactivated at temperatures of 800°C [20]. This temperature is high enough to pyrolize PFOA, thus it may decompose during GAC reactivation [21]. However due to lack of information in the literature, it is difficult to predict the degree of PFOA mineralization and the formation of side products.

The strong anion exchange resin (Table 1) seems to be a good adsorbent for PFOA since K_C values are higher than 100 (mg C/g) (mg C/l)$^{-1}$, whereas PFOS adsorption is substantially weaker (K_C: 49 (mg C/g) (mg C/l)$^{-1}$). The lower affinity of PFOS toward the ion exchange resin has been attributed to its slightly lower critical micelle concentration (CMC) and higher molecular volume compared to PFOA [18]. The latter factor probably dominates over CMC because the CMC of PFOA (8.7–10.5 mM) is in the same range as the CMC for PFOS (8 mM) [22, 23]. It has to be mentioned that the sorption isotherm for the ion exchange resin has been monitored in absence of other ionic species and competitive sorption has to be taken into account in real-water systems.

The sorption of surfactants in aqueous systems is very complex and hardly any mechanistic information is available about the processes contributing to adsorption of polyfluorinated compounds. Once adsorbed, the hydrophilic head group increases the negative charge of the surface leading to a stronger repulsion of equally charged compounds. Especially, ionic surfactants may lead to the formation of micelles on the surface of activated carbon which could lead to pore blocking. On the other hand, micelles may serve as an additional sorption layer which could increase the sorption efficiency.

In water treatment, competition for sorption sites may occur in presence of substances with equal or higher affinity to the sorbent such as natural organic matter in case of activated carbon applications. Inorganic ions may act as competitors in ion exchange filtration, whereas they could serve as sorption promoters for activated carbon filtration due to the reduction of the surface charge of the carbon [24, 25].

2.2 Membrane Filtration

Membranes are physical and in some cases chemical barriers to remove particles and solutes from a fluid. In water treatment, four types of membranes are applied: microfiltration (MF), ultrafiltration (UF), nanofiltration (NF), and reverse osmosis (RO). MF and UF membranes have an average pore size in the range of 0.1–10 μm (MF) and 0.002–0.1 μm (UF) and are used to remove particles from water. The pore size distribution of NF is 0.0005–0.002 μm and of RO below 0.0005 μm. However, as the membranes have usually a more or less broad pore size distribution, the borders between the different processes are somewhat blurred. The retardation effect of NF and RO is a combination of physical separation and chemical interactions of solutes with the membrane material. These membranes are used for the removal of dissolved compounds (e.g., pollutant control or reduction of hardness), which accumulate in the membrane concentrate [26]. Depending on the respective pore size, NF and RO membranes should in principle be suitable to remove PFC from process or drinking water. The rejection of PFOS, PFOA, perfluorobutanoic acid (PFBA), and PFBS treated with NF and RO membranes has been found to be >90% during a period of up to 4 days. Thereby, RO has been superior over NF with regard to rejection efficiency [27, 28]. For four different types of RO membranes, ≥99% rejection of PFOS has been achieved over a wide range of feed concentrations of 1–1,000 mg l^{-1} PFOS. The passage of up to 1% PFOS through RO membranes can be explained by diffusion through the polyamide separation layer. This has been confirmed by the finding of fluorine inside the separation layer [28]. The cross section of nonbranched PFOS is about 0.4 nm [18, 29], which is in the range of the pore sizes of NF and RO membranes, and thus might contribute to the migration effect. It has been indicated that PFOS may also lead to membrane fouling, resulting in flux or pressure loss especially for membranes with high initial fluxes. From an operational point of view, it has been recommended that high flux RO membranes should be avoided when treating water with high concentrations of PFOS (>30 mg l^{-1} PFOS) [4], because these membranes normally have a lower rejection effect than tighter membranes and the advantage of a high flux cannot be maintained for a long time. In the semiconductor industry, high concentrations of PFOS are sometimes coupled with the addition of a co-solvent (e.g., 2-propanol) to keep these chemicals in solution. This may adversely affect the membrane performance due to the increase of osmotic pressure [4]. For such cases, a pretreatment to remove the solvent prior to RO may be necessary.

3 Oxidative Chemical and Physicochemical Treatment

3.1 *Incineration*

Incineration is very efficient to destroy solid wastes. For water treatment, this technique may become feasible with prior sorption of the pollutants on an appropriate sorbent or after concentrating the pollutants, e.g., via membrane filtration. However, energy is lost by heating the sorbent or the water. PFOA is pyrolized when exposed to temperatures of >300°C [21]. The thermal degradation of this compound in the gas phase is enhanced when it exists as a salt (counter ions such as: Na^+, NH_4^+). Thereby, surfaces may act as a source for counter ions as has been demonstrated for crashed borosilicate glass [21]. During incineration, target compounds are heated to 1,000°C for at least 2 s. Under these conditions, the probably most recalcitrant fluoro chemical CF_4 is destroyed to >99% and can be considered to be incinerable [30]. However, the reaction pathways during incineration can be rather complex and product formation may be difficult to predict. Little is known about fluorinated products of incomplete combustion and their effect and behavior after release into the environment. More information about the thermodynamics, kinetics, and mechanisms of thermolysis and combustion of fluorinated compounds can be found in Refs. [13, 30–32]. Opportunities for optimization may be given by the addition of cations to transform PFC into less stable forms (e.g., metal salts) or addition of reactive agents like persulfate, which may contribute to the mineralization process via thermally generated sulfate radicals (Sect. 3.4.1).

3.2 *Sonolysis*

Sonolysis is based on expanding and compressing gas bubbles produced by ultrasound with frequencies of 20–1,000 kHz. The ultrasound can be generated at the bottom of a vessel and is reflected at the gas–water interface leading to standing waves. Small bubbles accumulate in areas of maximum amplitude where they oscillate. During expansion, gas is drawn into the bubbles that are heated up during the subsequent compression phase. In water, temperatures up to 4,600 K are reached leading to a decomposition of the water vapor inside the bubble. Thereby, OH radicals are formed at high concentrations on the bubble surface (10^{-2} M) [33]. Pollutants can be degraded by direct pyrolysis and indirectly via OH radicals. Perfluorinated compounds do not react with OH radicals and thus are probably degraded via pyrolysis only. Hydrophobic compounds accumulate at the gas/water interface and are more efficiently degraded than hydrophilic compounds [34]. Thus, the surface activity of PFOA and PFOS may be of advantage for sonolytically driven decomposition. PFOA and PFOS have been degraded with sonolysis (aqueous argon saturated solution; $t_{½}$ (PFOA) 22 min, $t_{½}$ (PFOS) 43 min; aqueous argon saturated solution: $t_{½}$ (PFOA) 45 min, $t_{½}$ (PFOS) 102 min; initial concentration of PFOA 24 μM and PFOS 20 μM; initial pH 4.8; 200 kHz, 3 W cm^{-2} [35]). After 60 min of sonolysis, perfluorinated carboxylic acids with chain length between C_1 and C_8 have been detected in the reaction solution for both PFOS and PFOA. Surprisingly, the sonolysis of PFOS has been reported to lead also to the formation of C_7 and C_6 perfluorinated sulfonates [36]. For investigating matrix effects, the sonolytical degradation of PFOA and PFOS has also been monitored in a groundwater [36]. This water contained volatile organic compounds (VOC) like acetone (0.12 mM), 2-propanol (0.041 mM), and diisopropyl ether (0.034 mM). The TOC concentration of the water was 20 mgC l^{-1}. To investigate the degradation of PFC via sonolysis, this water had been spiked with 100 μg l^{-1} PFOA and PFOS (250 W l^{-1}, 354 kHz, average energy transfer 72% based on calorimetric measurements, 10°C, Argon atmosphere). In the same study, a reference experiment had been conducted in pure water. The degradation rate for both compounds was significantly reduced in groundwater compared to the pure water matrix ($t_{½}$(Groundwater) 73 min; $t_{½}$(pure water) 29 min). The authors found that the addition of humic and fulvic acids with a concentration of 15 mg l^{-1} had no effect on the degradation of PFOA and PFOS and explained the detrimental effect of the groundwater with the presence of volatile compounds. They argued that these compounds evaporate into the cavitation bubble leading to a reduction of the temperature during its collapsing event due to their endothermic decomposition. However, this explanation is questionable since the authors also reported that acetone and 2-propanol with a concentration of 0.1–1 mM had no significant effect on the degradation of PFC in the pure water system. This is in agreement with a study of Rae et al. (2004) [37]. In this work, the effect of different alcohols (methanol, ethanol, *n*-propanol, *n*-butanol, *n*-pentanol, and *t*-butanol) on the temperature inside the cavitation bubbles has been investigated. Thereby, approximately 50 mM of the alcohols have been needed to reduce the temperature of the cavitation bubbles from 4,600 to 4,000 K [37].

3.3 Advanced Oxidation

One of the most reactive oxidants in water treatment is the OH radical ($^{\bullet}OH$) which is produced in advanced oxidation processes (AOP) such as ozone-based processes (ozonation, peroxon process (O_3/H_2O_2)) or UV-based processes (UV/H_2O_2, UV/O_3 or UV/TiO_2) as well as sonolysis of water [33]. The preferred reaction pathways are addition to C=C and C=N double bonds, hydrogen abstraction from C–H bonds, and in few cases electron transfer reactions [37]. Perfluorinated surfactants, however, do not exhibit the preferred reactive sites for OH radicals. The abstraction of fluorine from a carbon atom is thermodynamically unfavorable because the F–OH bond has a lower energy than the C–F bond (bond dissoziation energies: HO-F 216 kJ mol^{-1}, CF_3F 552 kJ mol^{-1}, R-CF_2-F 352 kJ mol^{-1}, R,R′-CF-F 508 kJ mol^{-1} [32]). Furthermore, the electron density of the ionic head group (e.g., carboxylates and sulfonates) is reduced by perfluorination hindering electron transfer reactions.

The kinetic constant for the reaction of trifluoroacetic acid with $^{\bullet}OH$ has been estimated to be $< 1 \times 10^6$ M^{-1} s^{-1} [38], indicating a low respectively no reactivity toward $^{\bullet}OH$. Hori et al. [39] have found that the addition of hydrogen peroxide to a UV application has a detrimental effect on the degradation of PFOA compared to direct UV-photolysis (radiation source: Xenon-doped mercury lamp), which supports the persistence of PFCs in presence of $^{\bullet}OH$. PFOS persists treatment with different methods of advanced oxidation (Fenton, UV/H_2O_2, ozonation in alkaline solution, and Peroxon process (O_3/H_2O_2)) over a time span of 120 min at room temperature [16]. Thus, advanced oxidation can be considered to be ineffective for the degradation of PFC especially in real-water systems such as surface water or wastewater, where competing reactions will strongly dominate over PFC degradation. However, partly fluorinated compounds may be degraded under conditions of advanced oxidation whereby the perfluorinated moiety might be released, which has been shown for 2-perfluoroalkyl-ethanol polyglycolether and *N*-ethyl-*N*-(heptadecafluoro-octane)-sulfonylglycinic acid [16].

3.4 *Alternative Oxidation Systems*

Some oxidation systems have been reported to decompose perfluorocarboxylic acids (PFA) and the corresponding sulfonic acids (PFS) at bench scale. The primary products are PFAs with shorter chain length, CO_2 and fluoride. The reaction is often proposed to be initiated by electron transfer from the ionic head group to an appropriate electron acceptor. In that regard, especially sulfate radical anions and electrolysis using boron-doped diamond electrodes have been reported to degrade PFA respectively PFS.

3.4.1 Sulfate Radical Anions

Sulfate radical anions ($SO_4^{\bullet -}$) are strong oxidizing agents for which reduction potentials of 2.5–3.1 have been reported [40, 41]. These radicals can be generated in various ways such as UV-photolysis and reduction of peroxodisulfate ($S_2O_8^{2-}$) or peroxomonosulfate (HSO_5^{-}) by transition metals as well as thermolysis of peroxodisulfate ($T > 40°C$) [42, 43]. Sulfate radical anions react more selectively via electron transfer, whereas hydroxyl radicals react predominantly by addition to double bonds and H-abstraction. The higher electrophilicity of $SO_4^{\bullet -}$ may result in a stronger relationship between the reaction rate constants and the molecular structure of the target molecule as it has been reported for some aromatic compounds [44]. In contrast to the reaction with $^{\bullet}OH$, perfluorocarboxylic acids of chain length between C_2–C_{11} have been degraded by $SO_4^{\bullet -}$ in pure water systems (UV/$S_2O_8^{2-}$ and thermolysis of $S_2O_8^{2-}$) [39, 45–48]. A second-order rate constant has been determined for trifluoroacetic acid (1.6×10^4 M^{-1} s^{-1} [38]) and estimated for PFA with chain length of C_3 (1.4×10^4 M^{-1} s^{-1}) and

C_4 (1.3×10^4 M^{-1} s^{-1}) [48]. The rate constants for the reaction of $SO_4^{\bullet-}$ with short-chain PFA are very low. This may be partly counterbalanced by the longer lifetime of $SO_4^{\bullet-}$ in presence of persulfate compared to $^{\bullet}OH$ in presence of H_2O_2 ($k_{SO_4^{\bullet-},S_2O_8}$: 5.5×10^5 M^{-1} s^{-1} [49], $k_{\bullet OH,H_2O_2}$: 2.4×10^7 M^{-1} s^{-1})(NIST[50]). A reaction pathway has been proposed by Kutsuna and Hori [48] as follows:

$$CF_3(CF_2)_6COO^- + SO_4^{\bullet-} \rightarrow CF_3(CF_2)_6COO^{\bullet} + SO_4^{2-}, \quad (1)$$

$$CF_3(CF_2)_6COO^{\bullet} \rightarrow CF_3(CF_2)_5CF_2^{\bullet} + CO_2, \quad (2)$$

$$CF_3(CF_2)_5CF_2^{\bullet} + O_2 \rightarrow CF_3(CF_2)_5CF_2OO^{\bullet}, \quad (3)$$

$$2CF_3(CF_2)_5CFOO^{\bullet} \rightarrow CF_3(CF_2)_5CFOOOOCF(CF_2)_5CF_3, \quad (4)$$

$$CF_3(CF_2)_5CFOOOOCF(CF_2)_5CF_3 \rightarrow 2CF_3(CF_2)_5CF_2O^{\bullet} + O_2, \quad (5)$$

$$CF_3(CF_2)_5CF_2O^{\bullet} \rightarrow CF_3(CF_2)_4CF_2^{\bullet} + COF_2, \quad (6)$$

$$COF_2 + H_2O \rightarrow CO_2 + 2HF, \quad (7)$$

$$CF_3(CF_2)_5CF_2O^{\bullet} + HSO_4^- \rightarrow CF_3(CF_2)_5CF_2OH + SO_4^{\bullet-}, \quad (8)$$

$$CF_3(CF_2)_5CF_2OH \rightarrow CF_3(CF_2)_5COF + HF, \quad (9)$$

$$CF_3(CF_2)_5COF + H_2O \rightarrow CF_3(CF_2)_5COO^- + HF + H^+. \quad (10)$$

It can be deduced from this proposition that the degradation of PFOA is a stepwise elimination of –CF_2 units, leading to shorter-chain PFA which are subsequently oxidized by sulfate radical anions until complete mineralization to CO_2 and HF. This is supported by the experimental findings of arising and subsequent decomposition of short-chain PFA and a nearly 100% yield of fluoride per CF_2-unit degraded [46].

However, the kinetics of a sulfate radical-driven degradation of PFC is slow. Thus, their degradation during water treatment probably is very energy-demanding (e.g., for UV/$S_2O_8^{2-}$). Since the generation of $SO_4^{\bullet-}$ via reduction of $S_2O_8^{2-}$ and HSO_5^- with transition metals is hampered by the relatively high reactivity of $SO_4^{\bullet-}$ toward the reduced metal species (NIST [50]), it is doubtful if the degradation of PFC occurs under such conditions. The degradation of PFA by sulfate radicals arising from the thermolysis of $S_2O_8^{2-}$ has been reported (pure water, 80°C, 0.364 mM PFOA, 50 mM $S_2O_8^{2-}$, synthetic air saturated solution, half-life time: ca. 30 min; derived from [45]). This might be applicable for the treatment of process water which is heated during an industrial process.

3.4.2 Electrolysis

The use of boron-doped diamond electrodes (BDDE) has recently become of interest for water treatment. With a boron-doped diamond layer covering the supporting electrode, high overpotentials can be achieved and be used for the oxidation of pollutants. In general, a reaction can happen either via anodic oxidation or indirectly by production of •OH from water oxidation. With a potential of 2.5–4.2 V, PFOS has been degraded and sulfate, fluoride and traces of trifluoroacetic acids have been found as reaction products. To prevent loss of volatile compounds (e.g., trifluoroacetic acid and HF), experiments have been performed in a gastight flow-through reactor. There, 80% of the fluorine bound to the molecule has been released as fluoride. Beside traces of trifluoroacetic acid, no intermediates such as other short-chain perfluorinated carboxylic acids have been observed. This suggests that the oxidation of intermediates has nearly been completed at the electrode surface before they could re-enter the bulk solution. In a flow-through reactor, a first-order reaction rate of 0.13 min^{-1} ($t_{½}$: 5.3 min) has been measured for PFOS degradation (working and counter electrodes p-Silicon with boron doped diamond film, anodic electrode surface 25 cm^2, galvanostatic operation, closed loop system with total volume of 2.0 l, background electrolyte $NaClO_4$: 10 mM, current density: 20 mA cm^{-2}, potential: 3.2 V, T: 22°C) [51]. Because OH-radical reactions are not likely to contribute to PFOS degradation, direct electron transfer probably is the main reaction pathway. However, in electrolysis, by-product formation has to be taken into account such as chlorine formation and the subsequent formation of halogenated organics.

3.4.3 Photolysis

PFOA does not absorb significantly UV light above 240 nm. With shorter wavelength, the molar absorption coefficient increases up to a value of ca. 500 M^{-1} cm^{-1} for 200 nm and to ca. 2200 M^{-1} cm^{-1} for 190 nm (pure water, pH 3; own measurements). Thus, direct photolysis is most efficient at short-wavelength UV and vacuum UV radiation (VUV) (<200 nm). Hori et al. [39] reported a PFOA half-life of 24 h using a Xe–Hg radiation source (initial concentration of PFOA: 1.35 mM, 200 W Xe–Hg Lamp, 22 ml, 4.8 atm. of oxygen). This radiation source emitted UV radiation mainly in the wavelength range of 300–400 nm. Only a small peak appears at 254 nm (ca. 5–10% of the total emission). The primary photoproducts have been PFA with shorter chain length. Photolysis of PFOA with a 15 W low pressure mercury lamp (emission line 254 nm with a minor band at 185 nm) (LP–Hg(254 + 185)) as radiation source results in a much higher degradation rate with a half-life of 90 min (initial concentration of PFOA: 100 μM, reaction volume 0.8 l). Analogous experiments using a 15 W LP Hg-lamp with no emission of radiation at 185 nm lead to hardly any degradation of PFOA within 2 h. Thus, the degradation of PFOA is mainly driven by the 185 nm radiation of the

LP–Hg(254 + 185) lamp. The reaction was accompanied by the formation of fluoride and short-chain PFA [52]. The energy demand can be calculated to be around 28 Wh l^{-1} for 50% degradation of PFOA. The high energy demand can be explained with the VUV absorbance of the water. VUV leads to the photolysis of water producing $^{\bullet}OH$ and hydrogen atoms. Even for a low absorption coefficient of water at 185 nm (ε_{185nm}: 3.6 M^{-1} cm^{-1}) [33], the penetration depth of this radiation is short since 99% of the light is absorbed by water within ca. 100 μm. Thus, most of the energy emitted by the radiation source is probably lost for water photolysis, which renders this process inefficient for direct photooxidation of PFA or PFS in water treatment.

4 Reduction

Reductive dehalogenation is often used for remediation at sites contaminated with persistent halogenated pollutants. The reduction process leads to a higher oxidizability of formed transformation products driven by biological or chemical processes. Thus, dehalogenation can be understood as a pretreatment.

Some reduction reactions of PFC have been investigated which will be summarized shortly.

Elemental Iron (Fe(0)) is used as a reductant for remediation of natural water halocarbon contamination (reduction potential Fe(0): −0.447 V). PFOS and PFOA have been decomposed with Fe(0). To accelerate the reaction, the process can be carried out under subcritical water condition. The degradation of PFOA and PFOS has been monitored under such conditions within an argon atmosphere (T: 350°C, pressure: 200 atm.). In this extreme environment, PFOS degraded with a half-life of 45 min and with a fluoride yield of 55% after 6 h [53]. The *hydrated electron* is a very strong reductant. Compared to most other reactants (e.g., $SO_4^{\bullet-}$) it reacts relatively fast with perfluorinated carboxylic acids (CF_3COO^-, $C_3F_7COO^-$ and $C_7F_{15}COO^-$ $k = 10^{6-7}$ M^{-1} s^{-1} [54], $k < 2.6 \times 10^6$ M^{-1} s^{-1} [55]). Besides pulse radiolysis, hydrated electrons can be produced from iodide under UV light in a photochemical process. The degradation of PFOS has been observed in the UV/iodide system under argon atmosphere. However, oxygen and iodine react very fast with solvated electrons ($k_{e-,I3-}$ >2 × 10^{10} M^{-1} s^{-1}, k_{e-I2}:5 × 10^{10} M^{-1} s^{-1}, k_{e-O2}:1,88 × 10^{10} M^{-1} s^{-1}[55]), which suppresses the degradation of PFOS or PFOA [56].

UV-photolysis of 2-propanol under alkaline conditions (pH > 12) leads to the formation of an 2-hydroxyprop-2-yl radical with a reduction potential of −2.1 V, which led to slow decomposition of PFOA with a half-life of 17.8 h [57].

Vitamin B_{12} contains Co as a central atom. Embedded into the vitamin, it can exist in the oxidation states Co(I), Co(II), and Co(III). Co(III) can be reduced to Co(II) and Co(I) with titanium citrate. Thereby, it is converted to its active form being a strong reductant. Under such conditions PFOS has been decomposed. The elimination of PFOS by vitamin B_{12} has been slow, since 66% of a mixture of branched isomers of PFOS has been degraded after approximately 3 days (70°C and pH 9) [58].

In summary, the reductive decomposition of PFC in water is possible but often extreme conditions are necessary. The reactions tend to be very energy-consuming and sometimes lead to a dramatic change in the water quality (e.g., high pH or temperature). Furthermore, only little is known about adverse effects such as by-product formation.

5 Biological Degradation

Partly fluorinated compounds can be biologically degraded including a certain degree of defluorination. Difluoromethane sulfonic acid has been defluorinated completely when this compound served as a sulfur source for a *Pseudomonas* species under aerobic conditions. An additional carbon and nitrogen source has been crucial as growth factors, indicating that the fluorinated compound is not used as a carbon or energy source. Trifluoroethane sulfonic acid and *1H,1H,2H,2H*-perfluorooctane sulfonate (H4PFOS) have also been partly defluorinated. The latter led to the formation of several volatile polyfluorinated compounds which have not been further characterized. No degradation or defluorination of the perfluorinated compounds trifluoromethane sulfonic acid and perfluorooctane sulfonic acid has been observed [59].

The degradation of a ^{14}C-labeled 8-2-fluorotelomer alcohol (3-^{14}C-*1H, 1H, 2H, 2H*-perfluorodecanol) has been monitored in bench scale experiments using the microbial community of an aeration tank from an industrial wastewater treatment plant. Within these experiments, the formation of several metabolites has been observed including PFOA and perfluorohexanoic acid (PFHxA). It has to be noted that the degradation of 8-2-fluorotelomer alcohol probably is a combination of several pathways leading to a parallel formation of PFOA and to a minor extent of PFHxA. In contrast, the direct biological transformation of PFOA to PFHxA is unlikely. However, the rise of fluoride and ^{14}C containing CO_2 indicates an enzymatic pathway for defluorination and degradation of the perfluorinated carbon skeleton [60].

Similar experiments for the degradation of the 8-2-fluorotelomer alcohol on the basis of microbial communities of a domestic wastewater treatment plant revealed a different pattern of metabolite formation. There, no formation of perfluorinated carboxylic acids like PFOA or PFHxA has been observed [61] suggesting that in this case the biological community of the industrial wastewater may be more adapted to fluorinated chemicals as a potential substrate.

The biological degradation of 2(*N*-ethyl-perfluorooctanesulfonamido) ethyl alcohol (N-EtFOSE), a monomer of a surface protection polymer, revealed the formation of 2(*N*-ethyl-perfluorooctanesulfonamido) acetic acid (N-EtFOSAA) with a yield of 23% and to a minor extent of PFOS (5.3% of transformed N-EtFOSE) under aerobic conditions within 96 h on the basis of sludge from a domestic wastewater treatment plant [62].

During domestic wastewater treatment, the concentration of several PFA and PFS has either remained constant or has even increased during the treatment process. The latter phenomenon probably is due to biological transformation of certain precursor compounds such as N-EtFOSAA [63].

In summary, the reported data indicate that biological degradation of fluorotelomers happens to appear under conditions of domestic and industrial wastewater treatment including partly mineralization of the perfluorinated moiety of the molecules. However, perfluorinated carboxylates or sulfonates are likely to be formed. Since there is no evidence for a biological degradation of PFA and PFS, they need to be considered terminal products of biodegradation. However, the finding of enzymatically driven cleavage of the C–F bond is promising for developments in biotechnology regarding biological degradation of PFC in applications such as water treatment or remediation. To this end, a deeper insight into the mechanism of the biological degradation of fluorotelomers is required. This is of particular importance because the biological conversion of these compounds might happen during, water treatment and environmental processes.

6 Promising Approaches for Combined Treatment Techniques

6.1 PAC-MF/UF

In conventional applications, powdered activated carbon (PAC) is added to the bulk solution prior to filtration. Thereby, the hydraulic retention time between the point of dosage and the separation unit is considered to be the time period of the sorption process. The equilibrium loads of activated carbon (AC) are higher in AC filtration than in conventional application of PAC [64]. This is because in the filtration process the equilibrium load ($q_{\text{equilibrium, filtration}}$) corresponds to the solute concentration of the filter influx ($c_{\text{in,filtration}}$), whereas for "floating" coal in a reaction tank the equilibrium load ($q_{\text{equilibrium, reaction tank}}$) corresponds to the solute concentration after sorption $c_{\text{out, reaction tank}}$ (see Fig. 1)

In the context of PFC removal, the improvement of PAC filtration is of particular interest since sorption of PFC on GAC may be inefficient (see Sect. 2.1). The "filtration effect" for PAC can be achieved by its immobilization on an appropriate supporting material such as polystyrene balls (Haberer process) [66, 67] or porous polyurethane cylinders [68]. A relatively new process in drinking water and pool water treatment is the combination of PAC with membranes (MF/UF). A full-scale application of PAC/UF for water treatment is the CRISTAL© process (*C*ombined *R*eactors *I*ntegrating a *S*eparation by membranes and *T*reatment by *A*dsorption in *L*iquid), which is applied in Slovenia, France, and Switzerland for drinking water treatment [20]. Thereby, PAC is added to the raw water prior to filtration via UF-membranes operated in cross-flow mode [20]. The PAC-membrane process can be

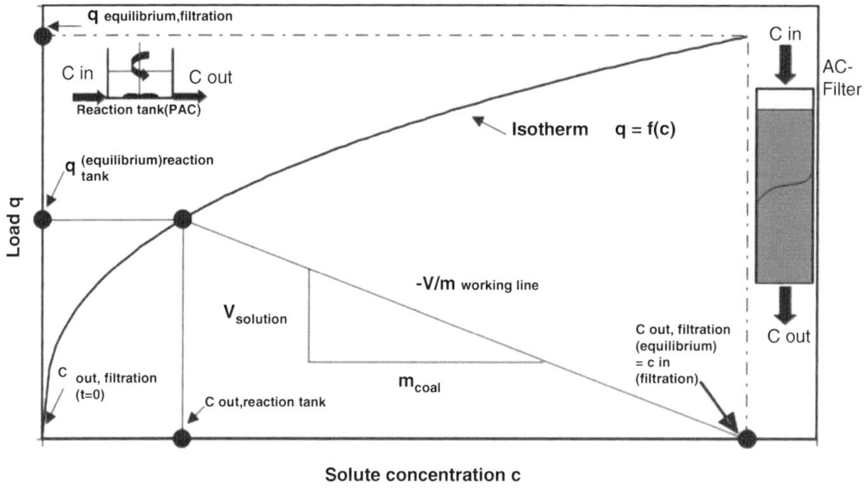

Fig. 1 The filtration effect; comparison of equilibrium loads of conventional powdered activated carbon treatment with activated carbon filtration; graphic reproduced from Ref. [65] with permission

further improved by dosage of submicron PAC with mean particle size of <1 μm. This has been shown for the removal of an odor compound (geosmin) from lake water (Sagami Lake, Japan). Hereby, a PAC/MF small-scale plant had been operated with PAC of different particle sizes (mean size 0.8 (PAC-0.8) and 10 μm (PAC-10)). A coagulant (poly-aluminum chloride) had been dosed between PAC dosage and the membrane filtration unit. Hereby, the PAC-0.8 was much more effective than PAC-10 with respect to geosmin removal. After a contact time of ca. 4 min > 98% of geosmin had been removed with PAC-0.8, whereas the PAC-10 led to a removal of < 62% (PAC-dose 2 mg l^{-1}, concentration of geosmin in raw water: ca. 514 ng l^{-1}). Under the same conditions, to achieve the high removal efficiency of PAC-0.8 required an increase in the PAC-10 dose to 20 mg l^{-1} [69]. Thus, 90% of PAC is saved by using the submicron powdered AC. Furthermore, the dose of PAC-0.8 led to a reduction of the transmembrane pressure compared to operation without PAC dosage. This can be explained by the adsorptive removal of organics which may cause membrane fouling [69]. The combination of submicron PAC with MF/UF is a promising advancement of conventional PAC treatment for the removal of PFC, because of improved sorption kinetics and maybe also equilibrium loads. Thus, smaller PAC reaction tanks can be used. An additional advantage of PAC/membrane filtration might also be the high flexibility to react on short-term episodes (e.g., accidental contamination of the raw water) by adjusting the PAC dose. Due to the lack of information about the behavior of PFC in the PAC/membrane system, research is promising within this context.

6.2 Ozone-AC

Ozonation prior to GAC filtration may lead to a better sorption performance for PFC. This may be due to a partial oxidation of the organic matter leading to a higher polarity and thus decreasing its sorptivity on AC surfaces [64]. This lowers the DOC-driven competition for active sites of the activated carbon and thereby increases the PFC removal efficiency. A synergistic effect of ozonation followed by AC filtration has been observed in a survey on the effectiveness of different tertiary treatment strategies in domestic wastewater treatment for micropollutant control. Compared with conventional ozonation, PAC treatment and sand filtration + activated carbon filtration, the ozone-GAC treatment led to the highest degree of PFOS removal of ca. 75% [70].

6.3 PAC-Activated Sludge

Activated sludge is a weak sorbent for PFOS [17] and the overall PFC removal efficiency in conventional wastewater treatment is low [63]. The process can be improved by the addition of PAC to the activated sludge which enhances the adsorptive removal of micropollutants including PFOA and PFOS [70]. Additional benefits are the reduction of the sludge index, improvement of the de-watering process and a higher calorific value. The load of pollutants bound to the sludge/activated carbon mixture might be decomposed during the advanced sludge treatment including anaerobic digestion and incineration. For wastewater treatment, PAC is typically added in dosages of $\leq$40 g m^{-3} [71, 72].

7 Conclusion

Separation techniques such as activated carbon treatment, ion exchange or membrane filtration coupled with incineration can effectively be used for PFC control in water treatment. However, a further characterization of the incineration process with regard to the formation of undesired by-products is necessary. Generally, the proposed processes for degradation of PFC often strongly influence the physical and chemical conditions of the water (change of pH, salt loads, high temperature) which would have to be re-adjusted for most points of application in water treatment. Furthermore, some of these techniques are very energy-consuming and in contrast to the discussed separation techniques, hardly any full scale experience exists. A comparison of the energy consumption for 50% transformation of PFOA and PFOS of different oxidative treatment methods is shown in Table 2. For comparison with conventional treatment methods, the energy demand of ozone production is included assuming a dose of 5 mg l^{-1} ozone (feed gas: air; ozone

Table 2 Comparison of different oxidative treatment strategies with respect to the energy demand for 50% degradation of PFOA resp. PFOS

Treatment	PFC	Initial PFC concentration (μM)	Experimental conditions	Energy demand ($Wh\ l^{-1}$) (50% degradation)/$t½$	Ref.
Sonolysis	PFOS	100	Ar, GW, 10°C	307/74 min	[36]
Sonolysis	PFOA	100	Ar, GW, 10°C	137/33 min	[36]
Sonolysis	PFOS	100	Ar, PW, 10°C	120/28 min	[36]
Sonolysis	PFOA	100	Ar, PW, 10°C	61/14 min	[36]
Photolysis	PFOA	100	N_2, PW, LP–Hg 254 + 185 nm, pH 3.7, 40°C	28/90 min	[52]
$UV/S_2O_8^{2-}$	PFOA	60	O_2, PW, LP–Hg 254 nm, 20°C	19/50 min	[75]
$UV/S_2O_8^{2-}$	PFOA	60	O_2, PW, LP–Hg 254 + 185 nm, 20°C	12/30 min	[75]
Electrolysis (BDDE)	PFOS	400	PW, pH 4–2.5, 22°C	≤0.2/5 min	[51]
Ozonation		–	–	*0.1/-*	–

The data is sorted by energy demand ($t½$: Half-life time of PFC, *PW* Pure water, *Ar, O_2, N_2* Experiment in argon, oxygen or nitrogen saturated solution, *GW* Ground water (landfill leachate), *LP* Low pressure Hg-Lamp, *BDDE* Boron-doped diamond electrodes)

concentration, gasphase: ca. 20 % by weight derived from [73]). This dose is very high with regard to drinking water treatment and also applied for wastewater ozonation [74].

It has to be mentioned that the experimental conditions of the different studies are not consistent (e.g., PFC concentration and pH) and the description of the experimental conditions is incomplete in some cases. Furthermore, the number of studies available seems to be somewhat limited. However, it might be useful to roughly asses the energy efficiency of the different treatment options. During sonolysis, both PFA and PFS are decomposed in pure water systems as well as in real-water matrices like landfill leachate water [36]. However, relative to the other treatment possibilities the energy demand appears to be high. Even under ideal conditions (pure water, argon-saturated solution), the energy demand derived from the presented studies is higher than that for direct UV-photolysis, $UV/S_2O_8^{2-}$, and electrolysis using BDDE. The data indicate that the presence of oxygen lead to a decrease of the energy efficiency. The addition of $S_2O_8^{2-}$ might further improve the sonolysis due to the formation of sulfate radicals from peroxodisulfate pyrolysis. The UV-based processes are a bit more energy efficient with regard to the degradation of PFA. The $UV/S_2O_8^{2-}$ process appears not to be substantially more efficient than direct photolysis using LP–Hg-Lamp emitting 254 + 185 nm. However, the photolysis of persulfate can also be achieved with ozone-free LP–Hg-lamps (no emission at 185 nm) which are safer in use. $UV/S_2O_8^{2-}$ has the additional advantage that existing UV/H_2O_2 plants can be retrofitted to a sulfate radical-based process by implementation of a $S_2O_8^{2-}$ dosage. The degradation of PFS by approaches based on sulfate radicals or UV-radiation has not been reported so far. Thus, these oxidation systems might be limited to the oxidation of fluorinated

Table 3 Potential of application of different treatment options for removal of PFC from water

Treatment method	PFC tested	Experience in operation and maintenance	Remarks
Membrane Filtration (NF, RO)	PFOA PFOS	State-of-the-art technique for water treatment	Good rejection of PFC, PFC-containing concentrates are produced
AC Treatment	PFOA PFOS PFBS	State-of-the-art technique for water treatment	Efficiency depends on the C-chain length (short chains are unfavorable), PFC-containing waste is produced
Incineration	PFC	State-of-the-art technique for waste treatment	Pre-enrichment is favorable (NF, UF, sorption), "off line" technique
Sonolysis	PFA PFS	Some industrial applications	Probably robust technique, no additional chemicals are needed, high energy demand
$UV/S_2O_8^{2-}$	PFA	Partly derivable from UV/H_2O_2 plants	Moderate to high energy demand, residual peroxodisulfate concentrations in the effluent have to be considered, acidification of the water is possible, hardly any information about real-water matrices and by-product formation is available
Photolysis	PFOA	Partly derivable from UV- and UV/H_2O_2 plants	High energy demand, hardly any information about real-water matrices available, no information about by-product formation
Electrolysis (BDDE)	PFOS	Some industrial applications	Low energy demand, hardly any information about application in real-water matrices and by-product formation available, high electrode surface is important due to limitation by diffusion; Fouling and scaling possible
Reductive treatment	PFOA PFOS	Some applications in remediation	Hardly any information about real-water matrices and by-product formation available, partly extreme changes of the physical and chemical status of the water (high pH, high temperature, etc.)

carboxylic acids. A surprising low energy demand for the oxidation of PFOS is achieved with electrolysis using BDDE. The energy demand seems to be comparable with the production of ozone for water treatment.

Table 3 summarizes the treatment options for the purification of PFC containing water. The different techniques are compared with respect to their potential for implementation into a water treatment process.

The removal of PFC in water treatment with state-of-the-art techniques is possible. Currently, activated carbon is applied for adsorption of PFC and in principle nanofiltration and reverse osmosis are suitable to reject PFC. However, the few Freundlich constants available indicate a low efficiency of the adsorption process especially in presence of natural organic matter considering that fractions of the DOC adsorb stronger than PFOA and PFOS (Sect. 2.1). Additional effort is needed for the treatment of the contaminated activated carbon and, in case of activated carbon filtration, to monitor the sorbent saturation and regular exchange of the sorbent embankment. A destructive treatment option could circumvent these problems. In case of the occurrence of PFA, the $UV/S_2O_8^{2-}$ process is interesting due to its analogy to the UV/H_2O_2 processes. Thus, some full scale experience is available with regard to the design and operation of the photochemical reactors. However, experience is lacking in the implementation of such reactors into a treatment chain, e.g., with respect to residual effluent concentration of $S_2O_8^{2-}$ that is its control and influence on subsequent treatment steps. The degradation of PFS can be achieved with ultrasound with a relatively high energy demand and BDDE. The degradation process during BDDE electrolysis is located directly at the surface of the anode. Therefore, a high surface area is needed to achieve an efficient process and efforts might be necessary to deal with scaling and fouling effects. Additional care has to be taken for by-product formation e.g. via anodic oxidation of chloride or bromide, which might yield halogenated organic compounds, chlorate and bromate for instance. The use of perchlorate as an background electrolyte (see Sect. 3.4.2) is probematically because this chemical is difficult to handle and it's contact to the waste or drinking water should be avoided.

Direct photolysis, $UV/S_2O_8^{2-}$, and electrolysis have been investigated as bench scale experiments in pure water matrices. However, the presence of dissolved organic and inorganic matter probably influences the energy efficiency and by-product formation pattern. Even though $UV/S_2O_8^{2-}$ and the use of BDDE seem to be promising techniques, the efficiency and product formation (such as chlorinated organics, chlorate and bromate) in real-water matrices at bench scale and pilot scale have to be carefully investigated to assess their practicability and safety for water treatment.

Furthermore hardly any information about the effect of micelle formation of PFC such as PFOA and PFOS during the discussed treatment options is available, since the corresponding experiments have been conducted with PFC concentrations far below their CMC.

The lack of knowledge about the behavior of fluoro chemicals in water treatment also includes incomplete understanding of their behavior in the environment, because some processes discussed as water or waste treatment techniques are related to chemical and physical processes in nature (combustion, UV-photolysis, formation of reactive species (e.g., ozone, hydroxyl radicals, and sulfate radical anions)) as well as biological processes. The knowledge about the fate of fluorinated chemicals in the environment can also contribute to improve treatment technologies.

Acknowledgments We would like to thank the Water Chemistry Society – Division of the German Chemical Society for their generous financial support. We also thank Prof. Dr. Clemens von Sonntag, Dr. Andriy Kuklya, and Dr. Klaus Kerpen for reviewing the manuscript and the fruitful discussions and Rani Bakkour for the measurement of the PFOA UV-spectra.

References

1. US-EPA (2009) Provisional health advisories for perfluorooctanoic acid (PFOA) and perfluorooctane sulfonate (PFOS). US Environmental Protection Agency
2. UBA (2006) Vorläufige Bewertung von perfluorierten Tensiden (PFT) im Trinkwasser am Beispiel ihrer Leitsubstanzen Perfluoroctansäure (PFOA) und Perfluoroctansulfonsäure (PFOS), Stellungnahme der Trinkwasserkommission des Bundesministeriums für Gesundheit (BMG) bei Umweltbundesamt vom 21.06.06, überarbeitet am 13.07.06
3. Fujii S, Polprasert C et al (2007) New POPs in the water environment: Distribution, bioaccumulation and treatment of perfluorinated compounds – a review paper. J Water Supply: Res Technol AQUA 56(5):313–326
4. Tang CY, Fu QS et al (2006) Use of reverse osmosis membranes to remove perfluorooctane sulfonate (PFOS) from semiconductor wastewater. Environ Sci Technol 40(23):7343–7349
5. Luvarà G, Pickering S, et al (2010) Risk evaluation industrial use of PFOA and PFOS Report of the RPS Group. Delft, Netherlands
6. Boulanger B, Vargo J et al (2004) Detection of perfluorooctane surfactants in great lakes water. Environ Sci Technol 38(15):4064–4070
7. De Silva AO, Mabury SA (2004) Isolating isomers of perfluorocarboxylates in polar bears (ursus maritimus) from two geographical locations. Environ Sci Technol 38(24):6538–6545
8. Ellis DA, Martin JW et al (2004) Degradation of fluorotelomer alcohols: A likely atmospheric source of perfluorinated carboxylic acids. Environ Sci Technol 38(12):3316–3321
9. Giesy JP, Kannan K (2001) Global distribution of perfluorooctane sulfonate in wildlife. Environ Sci Technol 35(7):1339–1342
10. Lange FT, Schmidt CK, et al (2007) Perfluorinated surfactants: The Perfluoroctanesulfonate (PFOS) substitude perfluorobutanesulfonate (PFBS) increasingly affects the raw water quality of rhine waterworks. Perfluorierte Tenside: Der PFOS (Perfluoroctansulfonat) – Ersatzstoff PFBS (Perfluorbutansulfonat) beeinflusst zunehmend die Rohwasser qualität von Rheinwasserwerken. 148(7–8):510–516
11. McLachlan MS, Holmstrom KE et al (2007) Riverine discharge of perfluorinated carboxylates from the european continent. Environ Sci Technol 41(21):7260–7265
12. Moriwaki H, Takata Y et al (2003) Concentrations of perfluorooctane sulfonate (PFOS) and perfluorooctanoic acid (PFOA) in vacuum cleaner dust collected in Japanese homes. J Environ Monit 5(5):753–757
13. Rayne S, Forest K (2009) Perfluoroalkyl sulfonic and carboxylic acids: A critical review of physicochemical properties, levels and patterns in waters and wastewaters, and treatment methods. J Environ Sci Health A Toxic/Hazar Subs Environmental Eng 44(12):1145–1199
14. Young CJ, Furdui VI et al (2007) Perfluorinated acids in arctic snow: New evidence for atmospheric formation. Environ Sci Technol 41(10):3455–3461
15. Rumsby PC, McLaughlin CL et al (2009) Perfluorooctane sulphonate and perfluorooctanoic acid in drinking and environmental waters. Philos Trans R Soc A: Math Phys Eng Sci 367(1904):4119–4136
16. [a] Schröder HF, Meesters RJW (2005) Stability of fluorinated surfactants in advanced oxidation processes - A follow up of degradation products using flow injection-mass spectrometry, liquid chromatography-mass spectrometry and liquid chromatography-multiple stage mass spectrometry. J Chromatogr A. 1082(1 SPEC. ISS.):110–119; [b] Takagi S, Adachi F et al. (2011) Fate of perfluorooctanesulfonate and perfluorooctanoate in drinking water treatment processes. Water Res 45(13):3925–3932
17. Ochoa-Herrera V, Sierra-Alvarez R (2008) Removal of perfluorinated surfactants by sorption onto granular activated carbon, zeolite and sludge. Chemosphere 72(10):1588–1593
18. Yu Q, Zhang R et al (2009) Sorption of perfluorooctane sulfonate and perfluorooctanoate on activated carbons and resin: Kinetic and isotherm study. Water Res 43(4):1150–1158
19. Sontheimer HF, Bernd Richard et al (1988) Activated carbon for water treatment. DVGW-Research centre University of Karlsruhe. ISBN: 3-922671-20-9

20. Rovel JM (2007) Water Treatment Handbook. Degremont-Verlag, Frankreich, ISBN: 978-1-84585-005-0
21. Krusic PJ, Marchione AA et al (2005) Gas-phase NMR studies of the thermolysis of perfluorooctanoic acid. J Fluorine Chem 126(11–12):1510–1516
22. Giesy JP and Kannan K (2002) Perfluorochemical surfactants in the environment. Environ Sci Technol 36(7)
23. Kissa E (2001) Fluorinated Surfactants and Repellents, 2nd edn. Marcel Dekker, New York
24. Schwunger MJ, Smolka HG (1977) Mixed adsorption of ionic and nonionic surfactants on active carbon. Colloid Polym Sci 255:589–594
25. Ihara Y (1992) Adsorption of anionic surfactants and related compounds from aqueous solution onto activated carbon and synthetic adsorbent. J Appl Polym Sci 44(10):1837–1840
26. Van Der Bruggen B, Vandecasteele C et al (2003) A review of pressure-driven membrane processes in wastewater treatment and drinking water production. Environ Prog 22(1):46–56
27. Panglisch S, Staben N, et al Verhalten perfluorierter Verbindungen bei der Trinkwasserproduktion. IWW Water Centre, unpublished
28. Tang CY, Fu QS et al (2007) Effect of flux (transmembrane pressure) and membrane properties on fouling and rejection of reverse osmosis and nanofiltration membranes treating perfluorooctane sulfonate containing wastewater. Environ Sci Technol 41(6):2008–2014
29. Carter KE, Farrell J (2010) Removal of perfluorooctane and perfluorobutane sulfonate from water via carbon adsorption and ion exchange. Sep Sci Technol 45(6):762–767
30. Vecitis CD, Park H et al (2009) Treatment technologies for aqueous perfluorooctanesulfonate (PFOS) and perfluorooctanoate (PFOA). Front Environ Sci Eng China 3(2):129–151
31. Burgess DR Jr, Zachariah MR et al (1995) Thermochemical and chemical kinetic data for fluorinated hydrocarbons. Prog Energy Combust Sci 21(6):453–529
32. Tsang W, Burgess DR Jr et al (1998) On the incinerability of highly fluorinated organic compounds. Combust Sci Technol 139(1):385–402
33. von Sonntag C (2008) Advanced oxidation processes: Mechanistic aspects, Wat Sci Technol (58):1015–1021
34. Henglein A, Kormann C (1985) Scavenging of OH radicals produced in the sonolysis of water. Int J Radiat Biol 48(2):251–258
35. Moriwaki H, Takagi Y et al (2005) Sonochemical decomposition of perfluorooctane sulfonate and perfluorooctanoic acid. Environ Sci Technol 39(9):3388–3392
36. Cheng J, Vecitis CD et al (2008) Sonochemical degradation of perfluorooctane sulfonate (PFOS) and perfluorooctanoate (PFOA) in landfill groundwater: Environmental matrix effects. Environ Sci Technol 42(21):8057–8063
37. Rae J, Ashokkumar M et al (2005) Estimation of ultrasound induced cavitation bubble temperatures in aqueous solutions. Ultrasonics Sonochem 12(5):325–329
38. Maruthamuthu P, Padmaja S et al (1995) Rate constants for some reactions of free radicals with haloacetates in aqueous solution. Int J Chem Kinet 27:605–612
39. Hori H, Hayakawa E et al (2004) Decomposition of environmentally persistent perfluorooctanoic acid in water by photochemical approaches. Environ Sci Technol 38(22):6118–6124
40. Neta P, Huie RE et al (1988) Rate constants for reactions of inorganic radicals in aqueous solution. J Phys Chem Ref Data 17(3):1027–1040
41. Wardman P (1989) Reduction potentials of one-electron couples involving free-radicals in aqueous-solution. 18:1637–1755
42. Liang C, Bruell CJ (2008) Thermally activated persulfate oxidation of trichloroethylene: Experimental investigation of reaction orders. Ind Eng Chem Res 47(9):2912–2918
43. Johnson RL, Tratnyek PG et al (2008) Persulfate persistence under thermal activation conditions. Environ Sci Technol 42(24):9350–9356
44. Neta P, Madhavan V et al (1977) Rate constants and mechanism of reaction of sulfate radical anion with aromatic compounds. J Am Chem Soc 99(1):163–164
45. Hori H, Nagaoka Y et al (2008) Efficient decomposition of perfluorocarboxylic acids and alternative fluorochemical surfactants in hot water. Environ Sci Technol 42(19):7438–7443

46. Hori H, Yamamoto A et al (2005) Efficient decomposition of environmentally persistent perfluorocarboxylic acids by use of persulfate as a photochemical oxidant. Environ Sci Technol 39(7):2383–2388
47. Hori H, Yamamoto A et al (2005) Efficient photochemical decomposition of long-chain perfluorocarboxylic acids by means of an aqueous/liquid CO_2 biphasic system. Environ Sci Technol 39(19):7692–7697
48. Kutsuna S, Hori H (2007) Rate constants for aqueous-phase reactions of SO4- with C2F5C(O) O- and C3F 7 C(O)O- at 298 K. Int J Chem Kinet 39(5):276–288
49. Yu XY, Bao ZC et al (2004) Free Radical Reactions Involving Cl., Cl2-., and SO4-. in the 248 nm Photolysis of Aqueous Solutions Containing S2O82- and Cl. J Phys Chem A 108(2): 295–308
50. NIST National institute of standards and technology. http://kinetics.nist.gov/solution/SearchForm
51. Carter KE, Farrell J (2008) Oxidative destruction of perfluorooctane sulfonate using boron-doped diamond film electrodes. Environ Sci Technol 42(16):6111–6115
52. Chen J, Zhang PY et al (2007) Photodegradation of perfluorooctanoic acid by 185 nm vacuum ultraviolet light. J Environ Sci 19(4):387–390
53. Hori H, Nagaoka Y et al (2008) Iron-induced decomposition of perfluorohexanesulfonate in sub- and supercritical water. Chemosphere 70(5):800–806
54. Huang L, Dong W et al (2007) Investigation of the reactivity of hydrated electron toward perfluorinated carboxylates by laser flash photolysis. Chem Phys Lett 436(1–3):124–128
55. Anbar M, Neta P (1967) A compilation of specific bimolecular rate constants for the reactions of hydrated electrons, hydrogen atoms and hydroxyl radicals with inorganic and organic compounds in aqueous solution. Int J Appl Radiat Isotopes 18(7):493–523
56. Park H, Vecitis CD et al (2009) Reductive defluorination of aqueous perfluorinated alkyl surfactants: Effects of ionic headgroup and chain length. J Phys Chem A 113(4):690–696
57. Yamamoto T, Noma Y et al (2007) Photodegradation of perfluorooctane sulfonate by UV irradiation in water and alkaline 2-propanol. Environ Sci Technol 41(16):5660–5665
58. Ochoa-Herrera V, Sierra-Alvarez R et al (2008) Reductive defluorination of perfluorooctane sulfonate. Environ Sci Technol 42(9):3260–3264
59. Key BD, Howell RD et al (1998) Defluorination of organofluorine sulfur compounds by Pseudomonas sp. strain D2. Environ Sci Technol 32(15):2283–2287
60. Wang N, Szostek B et al (2005) Fluorotelomer alcohol biodegradation - Direct evidence that perfluorinated carbon chains breakdown. Environ Sci Technol 39(19):7516–7528
61. Wang N, Szostek B et al (2005) Aerobic biotransformation of 14-C-labeled 8–2 telomer B alcohol by activated sludge from a domestic sewage treatment plant. Environ Sci Technol 39(2):531–538
62. Boulanger B, Vargo JD et al (2005) Evaluation of perfluorooctane surfactants in a wastewater treatment system and in a commercial surface protection product. Environ Sci Technol 39(15):5524–5530
63. Loganathan BG, Sajwan KS et al (2007) Perfluoroalkyl sulfonates and perfluorocarboxylates in two wastewater treatment facilities in Kentucky and Georgia. Water Res 41(20):4611–4620
64. Sontheimer H, Frick BR, et al (1985) Adsorptionsverfahren zur Wasseraufbereitung. DVGW-Research centre University of Karlsruhe. ISBN: 3-922671-11-X
65. Hobby R (2008) Kombination der UF/MF mit der Dosierung pulverförmiger Aktivkohle. unpublished document
66. Haberer K, Normann-Schmidt S (1991) Haberer processes. Combining contact flocculation, filtration, and PAC adsorption. J Am Water Works Assoc 83(9):82–89
67. Stukenberg JR, Hesby JC (1991) Pilot testing the Harberer process in the United States. J Am Water Works Assoc 83(9):90–96
68. Hobby R (1995) Entfernung organischer Störstoffe im Spurenbereich mit pulverförmiger Aktivkohle. Dissertation Universität Duisburg-Essen

69. Matsui Y, Aizawa T et al (2007) Adsorptive removal of geosmin by ceramic membrane filtration with super-powdered activated carbon. J Water Supply: Res Technol – AQUA 56 (6–7):411–418
70. Pinnekamp J and Merkel W (2008) Senkung des Anteils organischer Spurenstoffe in der Ruhr durch zusätzliche Behandlungsstufen auf kommunalen Kläranlagen – Gütebetrachtungen und Senkung des Anteils organischer Spurenstoffe in der Ruhr durch zusätzliche Behandlungsstufen auf kommunalen Kläranlagen – Kostenbetrachtungen. ISA RWTH-Aachen and IWW Water Centre Mülheim a.d.Ruhr
71. Cornel P (2010) Aktivkohle zur Abwasserbehandlung. Oral presentation at "Aktivkohletage Mannheim" 23./24. June 2010 Mannheim, Germany
72. Hollender J, Zimmermann SG et al (2009) Elimination of organic micropollutants in a municipal wastewater treatment plant upgraded with a full-scale post-ozonation followed by sand filtration. Environ Sci Technol 43(20):7862–7869
73. Langlais B, Reckhow DA, Brink DR (1991) Ozone in water treatment - Application and Engineering. AWWA research foundation and Lewis publishers ISBN: 0-87371-474-1. 2
74. Nöthe T, Fahlenkamp H et al (2009) Ozonation of wastewater: Rate of ozone consumption and hydroxyl radical yield. Environ Sci Technol 43(15):5990–5995
75. Chen J, Zhang P (2006) Photodegradation of perfluorooctanoic acid in water under irradiation of 254 nm and 185 nm light by use of persulfate. Water Sci Technol 54(11–12):317–325

Perfluorinated Compounds in Food

Marinella Farré, Marta Llorca, Sandra Pérez, and Damià Barceló

Abstract Per- and polyfluorinated compounds (PFC) are resistant to breakdown, are ubiquitous environmental contaminants which persist and may bioaccumulate through the food chain. In the recent years, increasing number of papers report high levels of PFC in blood, tissues, and breast milk from both occupationally and non-occupationally exposed human populations. The most important exposure pathways of PFC for humans are thought to be intake of drinking water, food and inhalation of dust.

This chapter provides a comprehensive examination of the current knowledge of food contamination by PFC, with special attention to the fundamental role chemical analysis play in the evaluation of these compounds' sources, levels, and exposure and risk assessment.

Keywords Dietary intake • Food analysis • Liquid chromatography • Mass spectrometry • Per- and polyfluorinated compounds • Risk assessment

Contents

M. Farré (✉) • M. Llorca • S. Pérez
Department of Environmental Chemistry, IDAEA-CSIC, Jordi Girona 18-26, Barcelona 08034, Spain
e-mail: mfuqam@cid.csic.es

D. Barceló
Department of Environmental Chemistry, IDAEA-CSIC, Jordi Girona 18-26, Barcelona 08034, Spain

Catalan Institute for Water Research-ICRA, C/Emili Grahit, 101, Edifici H2O, Parc Científic i Tecnològic de la Universitat de Girona, 17003 Girona, Spain

T.P. Knepper and F.T. Lange (eds.), *Polyfluorinated Chemicals and Transformation Products*, Hdb Env Chem (2012) 17: 127–153,
DOI 10.1007/978-3-642-21872-9_7,

Abbreviations

APCI	Atmospheric pressure chemical ionization
API	Atmospheric pressure ionization
APPI	Atmospheric pressure photoionization
BS	Baltic Sea
EPI	Enhanced product ion
ESI	Electrospray ionization
FID	Flame ionization detection
FTOHs	Fluorotelomer alcohols
GC	Gas chromatography
HDPE	High density polyethylene
HRMS	High resolution mass spectrometry
LIT	Linear ion trap
LV	Lake Vättern
MRM	Multiple reaction monitoring
MTBE	Methyl tertiary butyl ether
NI	Negative ion
PFC	Per- and polyfluorinated compounds
PFOA	Perfluorooctanoic acid
PFOS	Perfluorooctane sulfonates
PFPeA	Perfluoropentanoic acid
PLE	Pressurized liquid extraction
POPs	Persistent organic pollutants
Q1/q2/Q3	Quadrupoles in triple quadrupole instruments (q2 represents the collision cell)
QqQ	Triple quadrupole instrument
SLE	Solid liquid extraction
SRM	Single reaction monitoring
TBA	Tetrabutyl ammonium
TDI	Tolerable daily intake

1 Introduction

Per- and polyfluorinated compounds (PFC) are widely used in industrial applications due their hydrophobic linear carbon chain attached to one or more hydrophilic head [1–4]. Because of the properties, PFC are physically, chemically and biological stable. These compounds have been manufactured for more than 60 years, and are released into the environment following production and use.

PFC are ubiquitous environmental contaminants which persist and may bioaccumulate through the food chain [5–7]. These compounds have been detected worldwide in sediments and biota [7–11]. In the recent years, an increasing number of papers report high levels of PFC in blood, tissues, and breast milk from both occupationally and non-occupationally exposed human populations [12–16]. The most important exposure pathways of polyfluorinated compounds for humans are thought to be intake of drinking water, food and inhalation of dust [17, 18]. Due to the widespread distribution, environmental degradation, and metabolism of the PFC released into the environment, a very complex exposure situation exists [19]. As a result, the relative contribution to human exposure from different routes or from a single source is not yet known. Because of their bioaccumulation [20–23] and potential health concerns including toxicity [24–27], and their possible contribution to cancer promotion [28–30], non-governmental organizations, national and international authorities have addressed the PFC problem by several pressure and legislative actions. The total production of PFOS has been significantly reduced from 2000 to 2005. One major fire-fighting foam manufacturer, 3M, abandoned production of PFOS in 2000. In February 2006, EEUU regulators reached a voluntary agreement with eight companies to phase-out the use of PFOA. Under the agreement, companies will reduce emissions of this compounds from their facilities and consumer products by 95% by 2010, and work toward eliminating sources of PFOA by no later than 2015. As the largest global manufacturer and supplier of fluorotelomers such as Capstone, DuPont also plans to adapt its entire product line by year-end 2010 to utilize short-chain chemistry because short-chain molecules cannot break down to PFOA in the environment.

Since PFOS was identified as a PBT chemical (persistent, bioaccumulative, toxic) in 2002, different countries have been working in order to restrict its use and marketing in the industry. Canada, EEUU, and Europe have established the hazard risk assessment of PFOS, and proposed it as a candidate in the Stockholm convention on persistent organic pollutants (POPs). As an example, in EU the marketing and use of PFOS began to be restricted from 2006. It is currently being discussed if PFOA should be incorporated to Council Directive 76/769/EC as dangerous substances, as PFOS. However, PFOS is still manufactured by Germany (20–60 tonnes in 2003) and Italy (<22 tonnes in 2003).

PFC are now included in different health programs in EEUU, Canada, and Europe. The EU provides a better assessment of the distribution, toxicity, and persistence of these compounds in humans and PFC are the target several projects of the VII European Research Framework Programme. During the last years, several reviews have been published on PFC that summarize the analytical strategies [31, 32], biological monitoring data [33, 34], and recent advances in toxicology and their mode of action [35]. However, data on levels of PFC in the human diet are rather scarce [13] and few studies, however, report the levels of PFC in human food such as vegetables, meat, and eggs.

This chapter provides a comprehensive examination of the current knowledge of food contamination by PFC, with special attention given to the fundamental role chemical analysis played in the evaluation of these compounds' sources, levels, and exposure and risk assessment.

2 Overview of Analytical Methods for the Analysis of Food

Table 1 shows a summary of analytical methods applied for the analysis of PFC in food.

2.1 *Storage and Conservation of Food Samples for PFC Analysis*

Storage and conservation of samples for PFC analysis presents some critical steps because losses or contamination of the samples can easily occur.

In order to avoid contamination, different measures have been suggested. For example, pre-cleaning of the bottles prior to sampling by rinsing with semi-polar solvents [51]. However, less attention has been paid to the potential losses during storage. The main causes of losses are the adsorption to sample containers, the volatilization of some PFC, or transformations due to inappropriate conservation. There have been controversies about whether and which PFC can absorb to glass surfaces [52, 53]. The partial adsorption to glass containers of high concentrations to standard solutions has been reported [54], but it is expected that this will not happen in real samples with more complex matrices [14]. On the other hand, some authors reported that polymeric container, such as polypropylene (PP) and high density ethylene (HDPE), can also partially adsorb long chain compounds, such as PFOS and PFOA [55]. Another cause of losses is volatilization that can affect some volatile compounds, such as fluorotelomer alcohols (FTOHs), during sampling, storage, and sample pretreatment. In order to minimize these losses, it has been recommended to avoid headspace in the sampling bottles [56]. Long-term conservation of the samples is a critical point. Most of the authors report freezing, refrigeration, solvents addition, or acidification combined with refrigeration to preserve the samples [57]. However, it has been shown that when pH decreases, PFC becomes increasingly associated with the available protons, and then PFC can be more easily adsorbed to the container's surface [58]. Szostek et al. [59] investigated the stability of FTOHs in water and water samples mixed with

Table 1 Summary of analytical methods applied to food matrices

Sample	Compounds	Extraction and clean-up	Stationary phase	Mobile phase	Detection	LOD–LOQ	Ref.
Seafood	PFOS PFHS PFUnDA PFDA PFNA PFOA PFHpA PFHxA	Ion-pair liquid extraction method	Betasil C18	2 mM ammonium acetate/ methanol	LC-(QqQ)-MS/MS	LOQ 250 ng/kg in fish	[36]
Fish and oysters	PFOA, PFOS and PFDA	C18-SPE	Betasil C18	1 mM ammonium acetate (pH 6.0) and a methanol	LC-(QqQ)-MS/MS	LOQs were between 0.5 and 6 ng/l in 250 ml of water sample, while 5–50 ng/g (dry weight) for biological tissue sample	[37]
Fish, mussel, amphipods, and algae	PFOS PFOSA PFOA PFHS	Ion-pair liquid extraction method	Betasil C18 column	2 mM ammonium acetate aqueous solution and methanol	LC-(QqQ)-MS/MS	LOQ were in the range of 0.2–2 ng/g	[38]
Cheese	PFHxSH, PFOS, PFOA, PFNA, PFDA, PFUnA, PFDoA, PFTedA, PFOSA	Fluoros liquid–liquid extraction (F-LLE) in a triphasic solvent system and Oasis WAX- SPE	Flurosep RP Octyl HPLC column	Aqueous ammonium formate (5 mM) and methanol	LC-(QqQ)-MS/MS	LOD 1 μg/kg	[39]
Fish	PFOS and PFOA	Ion-pair liquid extraction method	Phenomenex Gemini C18 column	10 mM ammonium acetate and methanol	LC-(QqQ)-MS/MS	LOD 19 and 270 ng/kg LOQ 63 and 910 ng/kg for PFOS and PFOA, respectively	[40]
Fillets of raw fish, meat, whole-grain bread, vegetables, fruits, cheese, sunflower oil	PFCAs: PFBA, PFPeA, PFOA, PFNA, PFDA, PFUdA, PFDoA, PFTrDA, PFTeDA. PFBS, PFHxS, PFOS	(1) LLE (20 ml Tetrahydrofuran:water 75:25 (v/v)) (2) SPE by Oasis WAX and Supelclean ENVI-carb	Fluorosep RP Octyl column	6.3 mM aqueous ammonium formate at pH 4 and methanol	LC-(QqQ)-MS/MS	The LOQs and LODs were 0.7 and 0.5 ng/ml for short-chain PF (PFBA, PFPeA and PFBS); 0.025 and 0.01 ng/ml for long-chain acids (PFDoA, PFTrA and PFTeA); and 0.05 and 0.02 ng/ml for all other compounds except PFOA and PFUdA	[41]

(continued)

Table 1 (continued)

Sample	Compounds	Extraction and clean-up	Stationary phase	Mobile phase	Detection	LOD–LOQ	Ref.
						(LOQ = 0.2 ng/ml, LOD = 0.12 ng/ml) and PFHxS (LOQ = 0.5 ng/ml, LOD = 0.3 ng/ml)	
Breast milk	PFOSA, Me-PFOSA-AcOH, Et-PFOSA-AcOH, PFHxS, PFOS, PFPeA, PFHxA PFHpA, PFNA, PFDeA, PFUA, PFDoA	Oasis HLP- SPE	Betasil C8 column	20 mM ammonium acetate (pH 4) in water and methanol	LC-(QqQ)-MS/MS	LOD 0.1–3.2 (mg/ml of milk)	[42]
Fish	PFHxS, PFOS, PFOSA, PFOA, PFOA, PFNA, PFDA, PFUnA, PFDoA, PFTrA, PFTeA, PFPeDA	(1) Extraction by ACN in an ultrasonic bath (2) Clean-up on graphitized carbon and acetic acid (3) 0.5 ml extract digested by ammonium acetate	Discovery HS C18 analytical column	45 mM ammonium acetate–methanol	HPLC-HRMS (PFCSs and PFOSA); HPLC-MS/MS (PFCAs)		[43]
Breast milk	PFHxS, PFHpS, PFOS, PFPeA, PFHxA, PFHpA, PFOA, PFNA, PFDA, PFUdA	Precipitation and SPE with Oasis WAX	BEH C18 column	2 mM ammonium acetate–methanol	LC-(QqQ)-MS/MS	LOD in ng/l PFHxS 0.69; PFHpS 3.77; PFOS 1.54; PFPeA 5.50 PFHxA 2.91; PFHpA 2.98 PFOA14.15; PFNA 5.46 PFDA 1.44; PFUdA 1.30	[44]
Fish (raw and cooked)	N-EtPFOSA, PFOSA, N, N-Et_2PFOSA, N-MePFOSA, N,N-Me_2PFOSA	(1) LLE [2:1 (v/v) hexane: acetone] (2) Concentrated sulfuric acid (to remove lipids) (3) Clean-up by silica gel column chromatography	–	–	GC-PCI-MS	LODs for individual PFAs ranged from 0.03 to 10 ng/g	[45]
Veal, pork, lamb, white fish, seafood, canned fish, blue fish, whole milk, semiskimmed	PFBS, PFHxS, PFOS, PFDS, PFHxA, PFHpA, PFOA,	(1) 2 ml 200 mM sodium hydroxide (1 g sample), 30 min; 10 ml MeOH, 30 min; 150 μl, 4 M HCl	C18 analytical column	Acetonitrile mobile phasea and 10 mM	LC-(QqQ)-MS/MS	LODs 0.001–0.65 ng/g of fresh weight	[18]

milk, dairy products (cheese, yogurt, creamy yogurt, cream caramel, custard), vegetables, pulses, cereals, fruits, oils and fats, and eggs.	PFNA, PFDA, PFUnA, PFDoA	(2) Centrifugation (3) SPE (Oasis WAX) (4) Eluted in PP tubes with 25 mg EnviCarb and 50 µl of glacial acetic. (5) Extract filtered (2 µm nylon filter)		ammonium acetate			
Breast milk	PFOS, PFOA	Acetonitrile digestion and vortexed and centrifuged	On line with a Reprosil-Pur C18 AQ and separate by Reprosil-Pur vasic C8	2 mM ammonium acetate aqueous solution and acetonitrile	LC-(QTRAP)-MS/MS	LOQ 20 ng/l for PFOS and 200 ng/l for PFOA	[46]
Vegetables, cheese, margarine, milk, bread, strawberry jam, pork, beef, chicken, egg, fish, canned mackerel, salmon, cod, cod liver	PFBS, PFHxS, PFOS, PFHxA, PFHpA, PFOA, PFNA, PFDA, PFUnA, PFDoA	(1) Alkaline digestion (200 mM NaOH in MeOH), LLE (2) SPE (Oasis WAX) (3) ENVI-carb	<0.06–310 pg/g	2 mM ammonium acetate aqueous solution and acetonitrile	UPLC-ESI-MS/MS		[47]
Meat, fish, fast food	PFHpA, PFOA, PFNA, PFDA, PFUnA, PFDoA, PFTeDA	MeOH digestion	Genesis C18 analytical column	5 mM ammonium formate in water and acetonitrile/methanol	LC-(QqQ)-MS/MS	LOD between 0.5 and 1 ng/g range. LOQ estimated as three times the LOD value.	[48]
Lake trout	PFHxS, PFOS, PFOSA, PFHpA, PFOA, PFNA, PFDA, PFUnA, PFDoA, PFTeA, PFDS	Ion-pairing liquid extraction and filtered	Genesis C18 column	0.01 mM ammonium acetate aqueous solution and methanol	LC-(QTRAP)-MS/MS	LOD between 1 and 16 pg/g	[49]
Surface water, fish and birds	PFOS, PFHS, PFBS, PFOA, PFOSA	Ion-pairing liquid extraction (fish and bird) Oasis HLB SPE cartridges (water)	Betasil C18 column	2 mM ammonium acetate aqueous solution and methanol	LC-(QqQ)-MS/MS	LOQ 0.8 ng/l (water), 1.5 ng/g (fish livers) and 7.5 ng/g (bird livers)	[50]

PFBS Perfluorobutane sulfonate, *PFHS* perfluorohexane sulfonate, *PFHxA* perfluorohexanoic acid, *PFHpA* perfluoroheptanoic acid, *PFNA* perfluorononanoic acid, *PFDA* perfluorodecanoic acid, and *PFUnDA* perfluoroundecanoic acid, *PFUnDA* perfluoroundecanoate, *PFDA* perfluorodecanoate, *PFHpA* perfluoroheptanoate and PFHxA, *PFTeDA* perfluorotetradecanoate, PFOS, PFNA, PFOA acetonitrile during the storage. They concluded that aqueous samples can safely be stored in the freezer in a glass vial and sealed with a septum lined with alumina foil. Finally, biodegradation and biotransformations should be prevented. While good results were obtained when conservations were conducted in the freezer or using combinations of solvent (such as acetonitrile) and freezing [60], the use of biological inhibitors, such as formalin was found to suppress the MS responses during analysis [61].

2.2 *Food Sample Pretreatment, Extraction and Clean-Up*

In Fig. 1 a general scheme is presented summarizing extraction and clean-up strategies for the analysis of PFC in food.

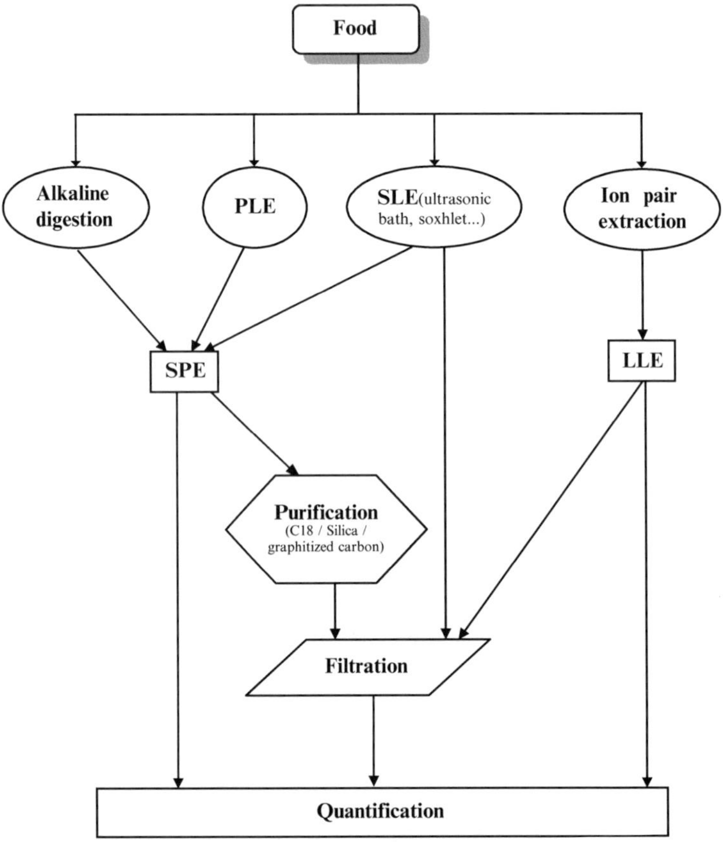

Fig. 1 Scheme of the extraction and cleanup methods

Main sample preparation and extraction procedures for the analysis of PFC in food have been based on:

- Ion pair extraction,
- Solid liquid extraction (SLE),
- Alkaline digestion, and
- Pressurized liquid extraction (PLE).

Ylinen et al. [62] developed an ion-pain extraction procedure employing tetrabutyl ammonium (TBA) counter ions for the determination of PFOA in plasma and urine in combination with gas chromatography (GC) flame ionization detection (FID). Later, Hansen et al. [53] improved the sensitivity of the ion-pair extraction approach using methyl tertiary butyl ether (MTBE) and by inclusion of a filtration step to remove solids from the extract, making it amenable for liquid chromatography coupled to tandem mass spectrometry (LC-MS/MS) determination. Ion pair extraction procedure has been the basis of several procedures for biota [63–65] and food samples [66, 67]. However, this method has shown to have some limitations, such as (1) co-extraction of lipids and other matrix constituents and the absence of a cleanup step to overcome the effects of matrix compounds and (2) a wide variety of recoveries are observed, typically ranging from <50% to >200%.

Solid liquid extraction (SLE) procedures have also been used by several authors [18, 19, 68, 69], using mixtures of hexane and acetone or using methanol. In general, extraction is followed by a cleanup step using sodium sulfate and acid attack to remove the lipid content. Tittlemier et al. [68] described this method for the determination of PFOSA and *N*-alkyl FOSAs in food, fish, and marine mammals. Homogenized samples were extracted by SLE with hexane:acetone (2:1), followed by a sample cleanup procedure. Extracts were dried over sodium sulfate, lipids were removed, and the extracts were passed through a silica gel column. In another study, the same group of researchers have [48] published a protocol to analyze PFOS and several PFCAs in food with LODs of 0.5–6 ng/g. In contrast to their method for neutral analytes described above, SLE was performed with MeOH. The resulting extracts were centrifuged and analyzed by LC/(-)ESI-MS/MS. 13C2-PFOA, 13C2-PFNA, 13C2-PFDA, and 13C4-PFOS were the isotope-labeled IS used in this study. Fromme et al. [28] reported a SLE procedure using ultrasonication and methanol. Extracts were cleanup using SPE with an anionic exchange cartridge.

Sample preparation by alkaline digestion has been also widely applied for the analysis of PFC in food. This procedure is based on digestion with sodium or potassium hydroxide in methanol followed by SPE. This procedure combined with SPE using Oasis-WAX cartridges has been applied for diverse foodstuff analyses. Vegetables, cheese, margarine, milk, bread, strawberry jam, pork, beef, chicken, egg, fish, canned mackerel, salmon, cod, and cod liver were also analyzed using alkaline digestion followed by SPE with Oasis-WAX by Haug et al. [47]. In another study, Jogsten et al. [70] used the alkaline digestion followed by SPE with Oasis-WAX for the analysis of a wide variety of foodstuff including raw, grilled and fried veal, pork and chicken, lamb liver, pate of pork liver, foi gras of duck, Frankfurt sausages, marinated salmon, lettuce, and common salt. Llorca et al. used this extraction procedure to study the PFC content in fish [71]and commercial baby food [16].

Modern extraction and cleanup techniques, such as pressurized liquid extraction, microwave-assisted extraction, or solid-phase microextraction have almost not yet applied to the analysis of PFC. Llorca et al. [71] reported the development and application of a PLE method for PFC determination in fish. This technique provided rapid and accurately clean extracts for sensitive analyses.

2.3 Qualitative and Quantitative Aspects of the Determination

Liquid chromatography-mass spectrometry (LC-MS) or liquid chromatography tandem-mass spectrometry (LC-MS/MS) has been in general the technique of choice for the analysis of PFC. Therein detailed information about the main experimental conditions used for analysis, such as LC-MS/MS precursor-product ion transitions were reported. Table 1 reports the main instrumental techniques based on mass spectrometry for the analysis of PFC in food.

Due to the complexity of the food samples, it is possible that the presence of some compounds in the matrix interferes with analyte determination. To date, this problem has been partially solved using LC–MS/MS. However, even when working in LC–MS/MS, certain compounds present in the sample can affect the initial ionization of the analyte through what is often called ion suppression or matrix effects.

ESI operating in the negative ion (NI) mode has been the interface most widely used for the analysis of anionic polyfluorinated surfactants. In addition, ESI has also been optimized for the determination of neutral compounds such as the sulfonamides PFOSA, Et-PFOSA, and t-Bu-PFOS. The use of atmospheric pressure photoionization (APPI) has been explored in few works [72–74]. Takino et al. [72] found the absence of matrix effects as the main advantage of this technology, but the limits of detection were considerably higher than those obtained by LC-ESI-MS/MS.

LC-MS/MS performed using triple quadrupole mass spectrometer (QqQ) combined with single reaction monitoring (SRM) is one of the more widely applied analyzer, in addition to be one of the better suited for quantification of PFC in food. Other analyzers used in the analysis of PFC in food samples by LC have been quadrupole-linear ion trap (QTrap) which usually allows LOQ lower than a QqQ, and high-resolution mass spectrometry (HRMS) for quantification and screening purposes. PFC contain carboxylic, sulfonic, hydroxyl, or sulfonamide group. They have acidic properties and can therefore dissociate. Therefore, electrospray ionization in the negative mode (ESI(-)) is best suited. LC-(ESI)-MS/MS is the technique most widely used in food analysis, allowing limits of detection in the pg–ng/g range. Recently, the analytical suitability of three different LC-MS/MS systems, QqQ), conventional 3D ion trap (IT), and quadrupole-linear IT (QqLIT), to determine trace levels of PFC in fish and shellfish were compared [75]. In this study, the accuracy was similar in the three systems, with recoveries always over 70%. Precision was better for the QqLIT and QqQ systems (7–15%) than for the IT system (10–17%). The QqLIT (working in SRM mode) and QqQ systems offered a linear dynamic range of at least three orders of magnitude, whereas that of the IT system was two orders of magnitude. The main advantage of QqLIT system is the high sensitivity, at least 20-fold higher than the QqQ system. Another advantage of QqLIT systems is the possibility to use enhanced product ion (EPI) mode and MS^3 modes in combination with MRM node for confirmatory purposes of target analytes in the complex matrices. These modes were applied in a recent investigation of breast milk samples and commercial baby food by Llorca et al. [16].

On the other hand, the gas chromatography (GC) is also used in the analysis of PFOSA and PFOSA isomers in food, coupled to a quadrupole mass spectrometer (Q).

3 Sources of Food Contamination

Two main sources of food contamination can be distinguished:

- Direct environmental exposure of plants and animals and/or bioaccumulation through the food chain, and
- Indirect contamination: Cooking, food packaging and food processes.

Direct environmental exposure of plants and animals and/or bioaccumulation through the food chain: There are several ways by which PFC can enter in drinking water and food. Food represents a part of the global environment which can be contaminated by chemicals such as PFC from many different sources. Following their release into the environment, PFC can enter plants and animals at the bottom of the food chain which are then consumed by animals higher up. Therefore, one of the main inputs of PFC in the food chain is the exposure of food-producing animals or plants to these compounds via environmental routes, i.e., exposures to contaminated air, water, or feed. Especially, contamination of the water cycle has been identified as one of the major causes of PFC in food. In addition, several studies report PFC contamination in drinking water [51, 76–79]. Removal efficiency of ionic PFC have been often been shown to be very limited [61]. Non-ionic PFC transform into the stable end products PFOS and PFOA. Hence, wastewater is one of the main influences of PFC in the water cycle. On the other hand, the use of sewage sludge as fertilizer and subsequent run-off was also found to contribute significantly to the contamination of surface, food, and drinking water [51].

In addition, bioaccumulation in food chains will lead to increased levels of PFC in animal-derived foods. Bioaccumulation of fish has been shown to be the main influences of PFC in dietary exposure. In a market basket study in Sweden, Berger et al. found that PFOS and PFOA concentrations were below the quantification limits in composite samples of foods of animal origin. However, predatory fish from the largest lake in Sweden had substantially elevated levels of several PFC.

In another work, Ericson et al. [18] studied the dietary exposure to PFC in Spain. In this study, the dietary intake of PFC was estimated for different age and gender groups and was found to be on average between 0.9 and 1.1 ng/kg bw/day for the adult male population. Fish, followed by dairy products and meats, were the main contributors to PFOS intake due to their bioaccumulation and biomagnification through the food chain. Similar conclusions were reported by Berger et al. [80]. In this work, fish consumption was identify as one of the main sources of human exposure in Sweden. Ostertag et al. [81] estimated the dietary exposure to PFC from traditional food among Inuit in northern Canada. In this study, the bioaccumulation of PFC through the food chain and their contribution to the Inuit dietary exposure were revealed. Recently, Haug et al. [82] explored the possible associations between concentrations of PFC in serum and sea food consumption, concluding that a significant relationship exists between estimated dietary intakes and serum concentrations.

Indirect contamination: Cooking, food packaging and food processes: Food preparation is another source of contamination [48], but preliminary data on the influence of domestic cookware on the levels of PFC in the preparation of food indicated no elevated levels for a limited number of experiments [83]. In addition, Del Gobbo et al. [45] reported that cooking decreases PFC concentrations in fish.

Packaging may also introduce chemicals into food, e.g., PFC used in greaseproof packaging for fast foods and special packaging. In these situations, PFC entry into food via migration from food package [48]. Fluorochemical-treated paper was tested to determine the amount of migration that occurs into foods and food-simulating liquids and the characteristics of the migration [84]. Additionally, microwave popcorn and chocolate spread were used to investigate migration. Results indicate that fluorochemical paper additives migrate to food during actual package use. For example, we found that microwave popcorn contained 3.2 fluorochemical mg/kg popcorn after popping and butter contained 0.1 mg/kg after 40 days at 4°C. Tests also indicate that common food-simulating liquids for migration testing and package material evaluation might not provide an accurate indication of the amount of fluorochemical that actually migrates to food. Tests show that oil containing small amounts of an emulsifier can significantly enhance migration of a fluorochemical from paper.

4 Food Contamination: Daily Intakes and Safety Limits

The characterization of health hazards of food contaminants, the assessment of the occurrence of undesirable compounds in food, and the estimation of the dietary intake are key issues in the risk assessment. In 2000, the European Commission published a White Paper on Food Safety, which underlined the importance of ensuring the highest possible standards of food safety and proposed a new approach to achieve them. Recently, PFC have gained increased scientific and socio-economic interest on the emerging environmental contaminants due to the unique combination of persistence, toxicity, and environmental prevalence. Risk assessment of the dietary exposure to PFC, however, is hampered by the lack of sufficient data about the occurrence of these contaminants in food.

A growing number of studies report on the occurrence of PFC in food. The outcome of these studies has been related to potential dietary intake and exposure levels (mainly by the estimation of the daily intake). It is important to remark that PFOS and PFOA tend to bind to certain proteins rather than bioconcentrate in fat, but they have also some potential to bioaccumulate in the food chain.

In the next sections, data published about the presence of PFC in drinking water and food will be revised. Special attention will be paid to fish contamination, since it has been well documented that PFC may bioaccumulate in fish and this accumulation tends to increase with increasing chain length [6, 85, 86]. Therefore, fish are an important dietary source of PFC for humans. In addition, a revision of daily intakes and safety limits is reported.

4.1 Fish Contamination

Among PFC fish contaminants, PFOS is the most crucial and prominent compound found. Reports suggest no considerable differences in PFC concentrations among freshwater and marine fish species. In contrast, the highest mean PFOS concentration (170 ng/g wet weight (wwt)) detected in lake trout, collected from Lake Ontario [80], from the Great Lakes ranged from 16 (Lake Michigan) to 121 ng/g wwt (Lake Erie) [49]. The PFOS concentration in lake trout from Lake Ontario increased significantly from 43 to 180 ng/g wwt in the period 1980–2001 [80]. However, this temporal trend was not confirmed by the study of Furdui, wherein an average PFOS concentration of 46 ng/g wwt was measured in 2001.

PFOA is the second most frequently detected PFC in fish but, it has been shown that PFOA is detected at much lower concentrations than is PFOS. Quantifiable concentrations of PFOA were detected in lake trout [49, 80], rainbow smelt, and alewife, with concentrations ranging from 0.16 to 6.8 ng/g wwt. The difference between the observed PFOS and PFOA concentrations in fish suggests a lower potential of PFOA to bioaccumulate in fish as compared to PFOS. This observation was further confirmed by laboratory experiments, which revealed a 1,000-fold lower bioconcentration factor for PFOA compared to PFOS [85, 87].

A restricted number of studies also reported other PFC and lower concentrations than PFOS were found. For example, Ye et al. [88] detected PFHxS at a maximum concentration of 1.89 ng/g wwt in a mixture of whole fish in the Missouri River, USA. Concentrations of the other PFC analyzed in this study were found in median concentration of 3.71 (PFHxA), 0.82 (PFDA), and 0.36 ng/g (PFHxS) wwt.

Martin et al. [80] detected relatively high mean concentrations of the longer chain PFC in fish collected from Lake Ontario, Canada. The highest concentration of these PFC was 8.3 ng/g wwt for PFUnA. These authors concluded that individual PFC were generally detected at lower concentrations than were PFOS, and total

PFOS equivalents (PFOS and PFOSA) exceeded the sum of all PFC by a factor of 1.8–12 within each species analyzed.

Tomy et al. [20] detected a relatively high mean concentration (92.8 ng/g wwt) of *N*-ethyl perfluorooctane sulfonamide in Arctic cod, ranging between 9.6 and 144.6 ng/g wwt. Since transformation of *N*-EtPFOSA to PFOS and PFOSA by rainbow trout microsomes has been reported [89], *N*-EtPFOSA is an important compound to measure in biota and in human samples.

Berger et al. [43] analyzed PFC in muscle tissue from edible fish species caught in the second largest freshwater lake in Sweden, Lake Vättern (LV), and in the brackish water Baltic Sea (BS). Again PFOS was the predominant PFAS found. PFOS concentrations were higher in LV (medians 2.9–12 ng/g fresh weight) than in BS fish (medians 1.0–2.5 ng/g fresh weight). Moreover, LV fish was more contaminated with several other PFAS than BS fish. This may be due to anthropogenic discharges from urban areas around LV. The PFAS pattern differed between LV and BS fish, indicating different sources of contamination for the two study areas. Human exposure to PFOS via fish intake was calculated for three study groups, based on consumption data from literature. The groups consisted of individuals that reported moderate or high consumption of BS fish or high consumption of LV fish, respectively. The results showed that PFOS intake strongly depended on individual fish consumption as well as the fish catchment area. Median PFOS intakes were estimated to be 0.15 and 0.62 ng/kg body weight (bw)/day for the consumers of moderate and high amounts of BS fish, respectively. For the group with high consumption of LV fish a median PFOS intake of 2.7 ng/kg bw/day was calculated. Fish consumption varied considerably within the consumer groups, with maximum PFOS intakes of 4.5 (BS fish) or 9.6 ng/kg bw/day (LV fish). These results suggested that fish from contaminated areas may be a significant source of dietary PFOS exposure. However, some controversial results were obtained by Nania et al. [90]. In this study the objective was to evaluate the contamination levels of PFOS and PFOA in edible fish of the Mediterranean Sea. Twenty-six fish muscles, 17 fish livers, 5 series of cephalopods (each composed of 10 specimens), and 13 series of bivalves (each composed of about 50 specimens) were used for the investigation. The results showed PFOA and PFOS levels in fishes and mollusks lower than those reported for analogue matrices in different geographic areas. According to their results no relation can be established between water contamination levels and posterior levels found in sea food. In another work, Llorca et al. [71] analyzed eight PFC in fish samples from Mediterranean Sea. The result of this study showed higher concentrations than those reported by Nania [90]. The results from Nania study also disagree with a recent study carried out under laboratory controlled conditions. Among the organisms studied, none of the bivalves accumulated PFC, and contrarily, insect larvae, followed by fish and crabs contained levels ranging from 0.23 to 144 ng/g ww of PFOS, from 0.14 to 4.3 ng/g ww of PFOA, and traces of PFNA and PFHxS.

In a recent study, fish consumption has been correlated. In this study carried out by Haug et al. [47], the possible associations between concentrations of PFC in serum and consumption of food with particular focus on seafood were studied, and estimated dietary intakes with determined serum PFC concentrations were compared. Concentrations of 19 PFC were determined in serum from 175 participants in the Norwegian Fish and Game Study and evaluated with respect to food consumption using multiple linear regression analysis. Associations between estimated individual total dietary intakes of PFC and serum concentrations were also explored. PFC concentrations in serum were significantly associated with the consumption of lean fish, fish liver, shrimps and meat, as well as age, breastfeeding history, and area of residence.

Although several authorities recommend not eating fish liver because of the risk associated with high intake of persistent organic pollutants (POPs), fish liver (and oil) is still consumed. it should be pointed out that PFC levels in liver are at least two orders of magnitude higher than that exists in the muscle tissue [90]. In Japan, concentrations of total PFC in skipjack tuna livers ranged from <1 to 83 ng/g wwt [91]. PFOS and PFUnA were the prominent compounds found.

Similar to fish, PFOS is the dominant PFC found in aquatic invertebrates such as shrimp, mussels, clams, and oysters [36, 92]. A few papers report on PFC levels in bivalves. Concentrations ranging from 1 to 6.0 ng/g wwt in oysters were reported from the Ariake Sea [93] and China [36]. Cunha et al. [94] measured high concentrations of PFOS in mussels from several estuaries in the North of Portugal. PFOS was detected in all the samples analyzed, and the concentrations were ranging 36.8 to 126.0 ng/g wwt. In a more recent work, Nania et al. [90] found higher PFOA than PFOS in clam but comparable levels were found in mussels, which was attributed to differences in habitat and feeding behavior.

Nowadays the bioaccumulation trends of PFC in aquatic organisms are not clear. In general, concentrations of PFC are expected to increase with increasing trophic level. This trend has been observed in the Great Lakes food chain [95]. However, higher concentrations of perfluoroalkyl contaminants were reported in lower trophic levels in seafood from China [36] and in invertebrate species from Lake Ontario [38]. However, there are some controversial results. It is clear that different processes including sorption processes to organic material, metabolic pathways are involved at the same time and data continue being inconsistent and the different sorption characteristics of the different types of PFC should be studied further.

Sorption coefficients of PFC are relatively low for C4–C8-carboxylic acids and increase with increasing chain length [96].

Biomagnification of PFOS in the estuarine food chain of the Western Scheldt estuary was observed by de Vos et al. [92]. On the other hand, it is not clear if there is a difference between the concentrations of PFC in edible fish from remote versus highly industrialized or urbanized areas. However, Gulkowska et al. [36] observed slightly higher PFOS concentrations in fish from the highly urbanized and industrialized areas.

More recently several authors studied the possible association between fish consumption and the levels of PFC in human blood. In recognition of the potential for human exposure to PFC via fish consumption, the Minnesota Department of Health has issued fish consumption advisories for contaminated sections of the Mississippi River (Minnesota Department of Health 2007). This advisory suggests that people limit their intake of fish to no more than one meal a week, if PFOS levels in fillet exceed 38 ng/g.

The provisional tolerable daily intake (TDI) values proposed by the European Food Safety Authority (EFSA 2008) and Health Protection Agency (HPA 2009) amount to 150 ng/kg body weight (bwt)/day and 300 ng/kg bwt/day, for PFOS and PFOA, respectively.

4.2 Foodstuff Studies

Studies that measure PFC in consumer food are limited. Table 2 presents a summary of concentration levels of PFC in foodstuff. One of the first studies was carried out in EEUU and was sponsored by 3M. The study measured PFOA, PFOS, and PFOSA in individual food samples including green beans, apples, pork, milk, chicken, eggs, bread, hot dogs, catfish, and ground beef [100]. Most samples had levels below the LOD (0.5 ng/g for all chemicals). The highest level of PFOA (2.35 ng/g ww) was detected in an apple purchased in Decatur, Alabama, the location of a 3M PFOA production plant. The highest level of PFOS (0.85 ng/g ww) was from milk purchased in Pensacola, Florida. Recently, in another study PFC among other POPs in composite food samples was evaluated from Dallas, Texas. The pattern of detection of PFC varied significantly in this study compared with the previous ones. In the previous studies, typically the most commonly detected PFC was PFOS, whereas in the study performed by Schecter [101], PFOS did not exceed the LOD, from 0.01 to 0.5 ng/g ww, in any samples, which is perhaps not surprising because it has been off the market since 2002. Instead, PFOA was found to exceed the LOD in 17 of 31 samples, with highest levels in butter (1.07 ng/g ww) and olive oil (1.8 ng/g ww). The relatively high levels of PFOA detected in the Schecter study might be attributed to the materials used in the processing and packaging of the food. Some food packaging materials contain trace amounts of PFOA, and PFC have been shown to migrate from packaging materials into food oils [84]. However, more research is required. A study of chemical contamination of food collected from 1992 to 2004 as part of the Canadian Total Diet Study examined PFC, including PFOS and PFOA [48]. Sampling continued through 2004, although PFOS was taken off the market in 2002. PFOA was detected at the highest levels in microwave popcorn (3.6 ng/g ww) and roast beef (2.6 ng/g ww), and PFOS was detected at the highest levels in beef steak (2.7 ng/g ww) and saltwater fish (2.6 ng/g ww). PFNA was detected in the beef steak sample (4.5 ng/g ww). LODs for PFC ranged from 0.4 to 5 ng/g ww.

Daily dietary intake of nine PFC, including PFOS and PFOA, were assessed in matched daily diet duplicates [13]. Diet samples were collected in year 2004 from 20 women in Osaka and Miyagi, Japan. Only PFOS and PFOA were detected in the diet samples without observing significant difference between cities. After adjusted by water content, diet concentration of PFOA was significantly higher in Osaka. The median daily intake calculated using the measured diet concentrations was

Table 2 Levels of PFC in food

Food	PFBS	PFHxA	PFHpA	PFHxS	THPFOS	PFOA	PFNA	PFOS	∑PFOSA	PFDA	PFUnDA	PFDS	PFDoDA	PFTDA	Ref.
Diet (ng/g)															
Bread	NR	NR	NR	NR	NA	<5	NR	<20	NA	NR	NR	NR	NR	NR	[97]
Miscellaneous cereals	NR	NR	NR	NR	NA	<5	NR	<10	NA	NR	NR	NR	NR	NR	[97]
Carcase meats	NR	NR	NR	NR	NA	<2	NR	<10	NA	NR	NR	NR	NR	NR	[97]
Offal	NR	NR	NR	NR	NA	<2	NR	<20	NA	NR	NR	NR	NR	NR	[97]
Meat products	NR	NR	NR	NR	NA	<2	NR	<10	NA	NR	NR	NR	NR	NR	[97]
Poultry	NR	NR	NR	NR	NA	<2	NR	<10	NA	NR	NR	NR	NR	NR	[97]
Fish	NR	NR	NR	NR	NA	<3	NR	<5	NA	NR	NR	NR	NR	NR	[97]
Oil and fats	NR	NR	NR	NR	NA	<1	NR	<0.5	NA	NR	NR	NR	NR	NR	[97]
Eggs	NR	NR	NR	NR	NA	<1	NR	1	NA	NR	NR	NR	NR	NR	[97]
Sugars and preserves	NR	NR	NR	NR	NA	<1	NR	1	NA	NR	NR	NR	NR	NR	[97]
Green vegetables	NR	NR	NR	NR	NA	<1	NR	<3	NA	NR	NR	NR	NR	NR	[97]
Potatoes	NR	NR	NR	NR	NA	1	NR	10	NA	NR	NR	NR	NR	NR	[97]
Other vegetables	NR	NR	NR	NR	NA	<10	NR	<3	NA	NR	NR	NR	NR	NR	[97]
Canned vegetables	NR	NR	NR	NR	NA	<5	NR	2	NA	NR	NR	NR	NR	NR	[97]
Fresh fruits	NR	NR	NR	NR	NA	<5	NR	<2	NA	NR	NR	NR	NR	NR	[97]
Fruit products	NR	NR	NR	NR	NA	<5	NR	<1	NA	NR	NR	NR	NR	NR	[97]
Beverages	NR	NR	NR	NR	NA	<0.5	NR	<0.5	NA	NR	NR	NR	NR	NR	[97]
Milk	NR	NR	NR	NR	NA	<0.5	NR	<0.5	NA	NR	NR	NR	NR	NR	[97]
Dairy products	NR	NR	NR	NR	NA	<5	NR	<2	NA	NR	NR	NR	NR	NR	[97]
Nuts	NR	NR	NR	NR	NA	<5	NR	<5	NA	NR	NR	NR	NR	NR	[97]
Beef stick	NA	NA	<0.6	NA	NA	<0.5	4.5	2.7	<LOD[a]	<1	<1	NA	<1	<3	[48, 98]
Roast beef	NA	NA	<0.6	NA	NA	2.6	<1	<0.6	<LOD[a]	<2	<2	NA	<1	<3	[48, 98]
Ground beef	NA	NA	<0.5	NA	NA	<0.4	<1	2.1	<LOD[a]	<1	<1	NA	<1	<3	[48, 98]
Luncheon meats, cold cuts	NA	NA	<0.4	NA	NA	<0.4	<1	0.5	<LOD[a]	<1	<1	NA	<1	<3	[48, 98]
Fish, marine	NA	NA	<0.4	NA	NA	<0.5	<1	2.6	<LOD[a]	<1	<1	NA	<0.8	<4	[48, 98]
Fish, freshwater	NA	NA	<0.4–1	NA	NA	<0.5–2	<1	1.5–2.0	<LOD[a]	<1–2	<1–2	NA	<0.9–2	<2–5	[48, 98]
Pizza	NA	NA	2.0	NA	NA	0.74	<1	<1	27.3[a]	<1	<1	NA	<1	<1	[48, 98]
Microwave pop corn	NA	NA	1.5	NA	NA	3.6	<1	0.98	15.3–18.9[a]	<1	<0.9	NA	<1	<1	[48, 98]

(continued)

Table 2 (continued)

Food	PFBS	PFHxA	PFHpA	PFHxS	THPFOS	PFOA	PFNA	PFOS	∑PFOSA	PFDA	PFUnDA	PFDS	PFDoDA	PFTDA	Ref.
Egg breakfast sandwich	NA	NA	<LOD	NA	NA	<LOD	<LOD	<LOD	11.9[a]	<LOD	<LOD	NA	<LOD	<LOD	[48, 98]
French fries	NA	NA	<LOD	NA	NA	<LOD	<LOD	<LOD	4.11–9.72[a]	<LOD	<LOD	NA	<LOD	<LOD	[48, 98]
Chicken nuggets	NA	NA	<LOD	NA	NA	<LOD	<LOD	<LOD	5.87[a]	<LOD	<LOD	NA	<LOD	<LOD	[48, 98]
Fish burger	NA	NA	<LOD	NA	NA	<LOD	<LOD	<LOD	3.82[a]	<LOD	<LOD	NA	<LOD	<LOD	[48, 98]
Vegetables	<LOD	<LOD	<0.004	<LOD	<LOD	<0.027	<LOD	0.022	<LOD	<LOD	<LOD	<LOD	<LOD	<LOD	[18]
Cereals	<LOD	<LOD	<0.009	<LOD	<LOD	<0.045	<LOD	<0.027	<LOD	<LOD	<LOD	<LOD	<LOD	<LOD	[18]
Pulse	<LOD	<LOD	<0.008	<LOD	<LOD	<0.080	<LOD	<0.069	<LOD	<LOD	<LOD	<LOD	<LOD	<LOD	[18]
White fish	<LOD	<LOD	<0.004	<LOD	<LOD	<0.065	<LOD	0.407	<LOD	<LOD	<LOD	<LOD	<LOD	<LOD	[18]
Sea food	<LOD	<LOD	<0.002	<LOD	<LOD	<0.029	<LOD	0.148	<LOD	<LOD	<LOD	<LOD	<LOD	<LOD	[18]
Tinned fish	<LOD	<LOD	<0.007	<LOD	<LOD	<0.126	<LOD	0.271	<LOD	<LOD	<LOD	<LOD	<LOD	<LOD	[18]
Blue fish	<LOD	<LOD	<0.010	<LOD	<LOD	<0.132	<LOD	0.654	<LOD	<LOD	<LOD	<LOD	<LOD	<LOD	[18]
Pork	<LOD	<LOD	<0.006	<LOD	<LOD	<0.053	<LOD	0.045	<LOD	<LOD	<LOD	<LOD	<LOD	<LOD	[18]
Chicken	<LOD	<LOD	<0.004	<LOD	<LOD	<0.004	<LOD	0.021	<LOD	<LOD	<LOD	<LOD	<LOD	<LOD	[18]
Veal	<LOD	<LOD	<0.003	<LOD	<LOD	<0.003	<LOD	0.028	<LOD	<LOD	<LOD	<LOD	<LOD	<LOD	[18]
Lamb	<LOD	<LOD	<0.012	<LOD	<LOD	<0.012	<LOD	0.040	<LOD	<LOD	<LOD	<LOD	<LOD	<LOD	[18]
Eggs	<LOD	<LOD	<0.005	<LOD	<LOD	<0.005	<LOD	0.082	<LOD	<LOD	<LOD	<LOD	<LOD	<LOD	[18]
Dairy products	<LOD	<LOD	<0.007	<LOD	<LOD	<0.007	<LOD	0.121	<LOD	<LOD	<LOD	<LOD	<LOD	<LOD	[18]
Whole milk	<LOD	<LOD	<0.015	<LOD	<LOD	<0.015	<LOD	<0.014	<LOD	<LOD	<LOD	<LOD	<LOD	<LOD	[18]
Semiskimmed milk	<LOD	<LOD	<0.004	<LOD	<LOD	<0.004	<LOD	<0.019	<LOD	<LOD	<LOD	<LOD	<LOD	<LOD	[18]
Fruits	<LOD	<LOD	<0.004	<LOD	<LOD	<0.004	<LOD	<0.017	<LOD	<LOD	<LOD	<LOD	<LOD	<LOD	[18]
Margarine	<LOD	<LOD	<0.014	<LOD	<LOD	<0.014	<LOD	<0.034	<LOD	<LOD	<LOD	<LOD	<LOD	<LOD	[18]
Oil	<LOD	<LOD	<0.035	<LOD	<LOD	<0.035	<LOD	<0.099	<LOD	<LOD	<LOD	<LOD	<LOD	<LOD	[18]
Milk and infant formula															
Infant formula	<0.7	NA	<1.2	<1.34–3.59	NA	<48.3	<2.20	<11.0–11.3	NA	NA	NA	NA	NA	NA	[99]
Dairy milk	<0.7	NA	<1.2	<1.34–3.82	NA	<48.3	<2.20	<11.0	NA	NA	NA	NA	NA	NA	[99]

LOQ not detected; *NA* not analyzed; *NR* not reported; *PFBS* perfluorobutane sulfonate; *PFDA* perfluorodecanoic acid; *PFDoDA* perfluorododecanoic acid; *PFDS* perfluorodecane sulfonate; *PFHpA* perfluoroheptanic acid; *PFHxA* perfluorohexanoic acid; *PFHxS* perfluorohexane sulfonate; *PFNA* perfluorononanoic acid; *PFOA* perfluorosulfonate acid; *PFOS* perfluoroctane sulfonate; ∑*PFOSA* perfluorooctanesulfonamide; *PFTDA* perfluorotetradecanoic acid; *PFUnA* perfluoroundecanoic acid; *THPFOS*: 1H,1H,2H,2H-perfluoroctanesulfonic acid

[a]Sum of *N*-ethyl-PFOSA, PFOSA, *N*,*N*-diethyl-PFOSA, *N*-methyl-PFOSA, and *N*,*N*′-dimethyl-PFOSA

1.47 ng PFOS/kg b.w. and 1.28 ng PFOA/kg b.w. for Osaka, and 1.08 ng PFOS/kg b.w. and 0.72 ng PFOA/kg b.w. for Miyagi.

In Europe, one of the first studies was carried out by the UK Food Standards Agency which published results of PFC analysis in food collected from the 2004 Total Diet Study [97]. PFOS exceeded the LOD in potatoes, canned vegetables, eggs, sugars, and preserves, with highest levels detected in potatoes (10 ng/g ww), including fresh potatoes as well as potato chips, french fries, and hash browns, whereas, PFOA was detected only in potatoes (1 ng/g ww).

In Germany, Fromme et al. [28] conducted a study to quantify the dietary intake of PFOS, PFOA, PFHxS, PFHxA, and PFOSA using 214 duplicate diet samples and to estimate individual intakes based on the blood levels of PFOS and PFOA. The median (90th percentile) daily dietary intake of PFOS and PFOA was 1.4 ng/kg b.w. (3.8 ng/kg b.w.) and 2.9 ng/kg b.w. (8.4 ng/kg b.w.), respectively. PFHxS and PFHxA were detected only in some samples above the detection limit with median (maximum) daily intakes of 2.0 ng/kg b.w. (4.0 ng/kg b.w.) and 4.3 ng/kg b.w. (9.2 ng/kg b.w.), respectively. PFOSA could not be detected above the limit of detection of 0.2 ng/g f.w indicating that this indirect route of exposure is of less significance.

Another study examined the dietary intake of PFC and estimated for various age/gender groups of the population of Tarragona County (Catalonia, Spain) [18] during 2006. PFC levels were determined in 36 composite samples of foodstuffs randomly purchased in various locations. PFOS, PFOA, and PFHpA were the only detected PFC in foodstuffs. The most commonly detected PFC was PFOS, in 24 of 36 samples, with the highest levels in an uncooked bluefish composite sample (0.654 ng/g ww), which included salmon, sardines, and tuna. PFOA was found only in whole milk, at relatively low levels (0.055 and 0.058 ng/g ww). On average, for a standard adult man (70 kg of body weight), the dietary intake of PFOS was estimated to be 62.5 or 74.2 ng/day (assuming ND $=$ 0 or ND $=$ 1/2 LOD, respectively). Fish, followed by dairy products and meats, were the main contributors to PFOS intake.

Several PFC have been detected in human blood from populations in North and South America, Asia, Australia, and Europe [102]. Different studies in Europe showed that highest PFOS concentrations were found in Poland, followed by Belgium, being comparable to Sweden, with lowest concentrations in Italy [14]. These results indicate differences in exposure across Europe. However, the sources and pathways of human exposure to PFC are currently not well understood [48]. The wide variety of industrial and consumer applications leads to numerous possibilities for release of PFC into the environment and subsequent exposures to humans via environmental routes and media. However, the relative uniform distribution of blood concentrations of PFC in children and the majority of adult populations points to a common major source, possibly food.

PFOS and PFOA chemicals have also been detected in human milk [16, 44, 91]. The mechanism by which polyfluorinated substances are transferred from blood to milk is not completely known even though it is related to the strong bound of PFC to proteins. An interesting study by Roper et al. [103] quantified PFOS and PFOA in

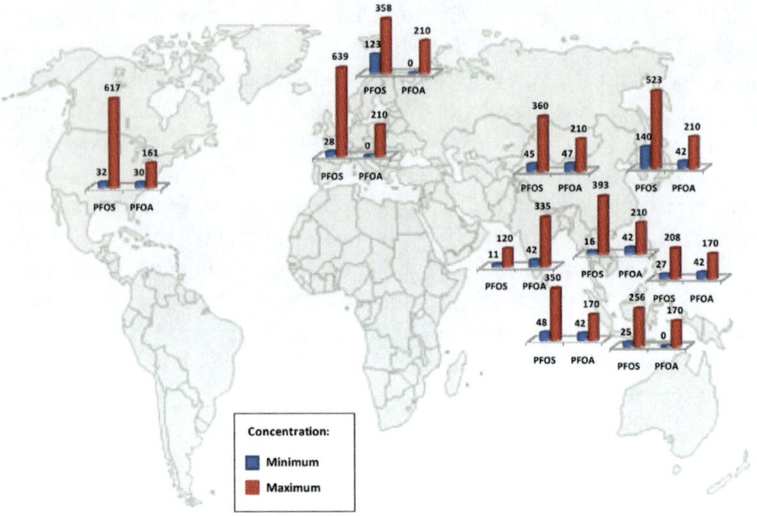

Fig. 2 Maximum and minimum concentrations (μg/ml) of PFOS and PFOA in breast milk from several countries. Data from Spain (Llorca et al.); German and Hungary (Völker et al.), Sweden (Karrman et al.), China (So et al.), Asian countries (Tao et al.) and EEUU (Tao et al.)

food emulsions produced by high-pressure homogenization, showing that both compounds, initially found in water, were distributed via protein binding to the creamed phase. Concentrations of PFC in human milk have been examined in a few studies, and the results for PFOS and PFOA in several countries are shown in Fig. 2. Lactation is a considerable source of exposure for infants, with levels of PFOS and PFOA in human breast milk ranging from 28 to 639 ng/l for PFOS and 0–210 ng/l for PFOA. In addition to PFOS and PFOA, PFPA (up to 1.56 μg/l), PFHxA (up to 0.82 μg/l), PFHxS (0.03–0.17 μg/l), and PFOSA (up to 0.03 μg/l), PFNA (up to 0.06 μg/l), PFDA (up to 0.02 μg/l) and PFUnDA (up to 0.06 μg/l) were less frequently detected [40, 41, 46, 76, 104, 105].

It must be noted that the international regulatory organizations (World Health Organization (WHO), European Union (EU)/EFSA, the US EPA, etc.) have not established safety limits yet for PFC in drinking water. However, recently, Schriks et al. [106] derived provisional drinking water guideline values for PFOS and PFOA of 0.5 and 5.3 μg/l, respectively, on the basis of the tolerable daily intake (TDI) values proposed by EFSA (2008).

The occurrence of PFC in surface and drinking waters of the Ruhr and Moehne area [51] in Germany caused a high concern, in view of the possible effects on humans and the environment. Immediately after detection of high concentrations of PFOA in drinking water the German Drinking Water Commission (DWC) of the German Ministry of Health at the Federal Environment Agency set for the first time in June 2006 a worldwide health-based guide value for safe lifelong exposure at 0.3 μg/l (sum of PFOA and PFOS). In addition, a set of measures were proposed and the local health authorities recommended that residents in parts of Arnsberg do not use tap water for preparation of baby food and advised pregnant women to avoid regular intake of such water. Additionally, recent EU regulations require phasing out use of PFOS and asked to voluntarily reduce the intake of PFOA. New and shorter chained PFC (C4–C7) and their mixtures are being introduced as replacements. These shorter chained compounds could be main contributors to total PFC levels in drinking water in future, especially because short-chained PFC are difficult to remove from drinking water by common treatment techniques and also by filtration over activated carbon. A recent study by Wilhelm et al. [78] provided a summary of the data from the regularly measured PFC levels in drinking water and in the drinking water resources in North Rhine-Westphalia for the sampling period 2008–2009 to give an overview of the general approach to assess PFC mixtures and to assess short-chained PFC using toxicokinetic instead of (sub) chronic data. The new approach to assess short-chained PFC is based on a ranking of their estimated half-lives for elimination from the human body. Accordingly, the authors considered the following provisional health-related indication values (HRIV) as safe in drinking water for lifelong exposure: PFBA 7 μg/l, PFPA 3 μg/l, PFHxA 1 μg/l, PFHpA 0.3 μg/l, PFBS) 3 μg/l, PFPS 1 μg/l, PFHxS 0.3 μg/l, and PFHpS 0.3 μg/l. For all PFC, the long-term lowest maximal quality goal (general precautionary value, PV_g) in drinking water was set to −0.1 μg/l.

It should be point out that most monitoring studies have focused only on PFOS and PFOA, but a few also reported on other PFC that appear at rather high concentrations in potable water such as PFBS, PFDoA, perfluoropentanoic acid (PFPeA), and PFHxA [51, 76, 79, 107]. Therefore, it is important to increase monitoring efforts with a view to setting more comprehensive safety limits for PFC in potable water.

In 2006, the EPA and the eight major PFC manufacturing companies in the industry launched the 2010/15 PFOA Stewardship Program, in which companies committed to reduce global facility emissions and product content of PFOA and related chemicals by 95% by 2010 and to work toward eliminating emissions and product content by 2015 (http://www.epa.gov/oppt/pfoa/pubs/stewardship/index.html).

Recently, in New Jersey, the Department of Environmental Protection developed preliminary health-based drinking water guidance for PFOA of 40 ng/l (http://www.defendinscience.org/case_studies/upload/pfoa_dwguidance.pdf).

Several scientific institutions have derived TDIs from toxicological end points by applying an uncertainty factor. The Scientific Panel on Contaminants in the Food Chain (CONTAM) established a TDI for PFOS of 150 and for PFOA of 1.5 μg/kg bwt/day (EFSA 2008). The UK Committee on Toxicity of Chemicals in Food, Consumer Products and the Environment (COT) proposed a TDI for PFOS and PFOA of, respectively, 300 and 3,000 ng/kg bwt/day (COT 2006a,b). Furthermore, the German Federal Institute for Risk Assessment proposed a TDI of 100 ng/kg bwt/day for both PFOS and PFOA (BfR 2006).

However, in addition to ingestion there are several routes of human exposure to PFC including mouthing of articles, dermal contact (i.e., during consumer use of articles containing PFC), or inhalation (air or indoor dusts) [104, 108]. Considering some the potential routes of human exposure to PFOS and PFOA, Fromme et al. [69] have estimated the overall mean and high daily intake for a non-occupationally exposed adult population. Similarly, Trudel et al. [108] reported a comprehensive assessment of consumer exposure to PFOS and PFOA from a variety of environmental and product-related sources. To identify the relevant pathways leading to consumer exposure to these compounds, scenario-based approach has been applied. The study shows that North American and European consumers are likely to experience ubiquitous and long-term uptake doses of PFOS and PFOA in the range of 3–220 ng per kg body weight per day (ng/kgbw/day) and 1–130 ng/kgbw/day, respectively. This study does not consider precursor compounds that could take up and convert to PFOS and PFOA within the human body.

The relative importance of metabolic transformation of precursor compounds in exposure to PFOS and PFOA has been scarcely evaluated and, to our knowledge, the only study that afforded the problem by a Scenario-Based Risk Assessment (SceBRA) approach estimated the relative importance of precursor-based doses of PFOS and PFOA of 2–5% and 2–8% in an intermediate scenario and 60–80% and 28–55% in a high-exposure scenario. This indicates that these precursors are of low importance for the general population.

PCF exposure risk assessment for infants and evaluation of lactation as an exposure pathway has also been assessed [91]. For a 5-kg Swedish child consuming breast milk at a rate of 800 ml/day, PFOS intake can be estimated at 48–380 (mean, 160) ng/day, or approximately 9.6–75 (mean, 32) ng/kg b.w. per day. The calculated total amount of PFC transferred by lactation to a breasted infant in this study was, approximately, 200 ng/day. As all the Swedish human milk samples came from the area of Uppsala, this intake estimate may not be representative of breastfed infant exposure to PFOS throughout Sweden. Likewise, the Swedish milk-based intake values may not in principle be extendable to the other European breastfed infants despite the concentrations detected in the Swedish human milk seem to be corroborated by the findings reported for other countries (Fig. 2).

In a recent study [16], for first time commercial baby food was evaluated and a estimation and the estimation of daily intake was also evaluated.

5 Conclusions and Future Trends

As in any analytical procedure, a suitable choice of sample-preparation technique is essential for accurate and reliable characterization of the PFC in food. However, because of the peculiarities of these compounds – especially, the background contamination problems (laboratory materials made of or containing perfluroethylene or perfluorolkyl compounds) that are a source of interferences for the analysis of PFC in trace or ultratrace concentrations – selection of the analyte-isolation and pre-concentration technique, as well as careful optimization of the corresponding operational parameters, is of paramount importance. An accurate and precise analysis of PFC in food is feasible if a number of decisive aspects are addressed.

Among emerging approaches, the potential of LC-MS/MS for high-throughput multi-analyte analysis ant its strong presence in future trends in PFC analysis is unquestionable. Perhaps the next frontier to be breached in this area will be the rapid screening and analysis using bioanalytical tools. The work reported until now has been performed by chemical analysis. To our knowledge, no biological technique has been developed for the determination of PFC.

On the evaluation of the dietary intake of PFC, few studies have investigated the occurrence of these compounds in different types of food. Thus, the values describing the occurrence of PFOS and PFOA in the human diet are still fraught with considerable uncertainty. Only a few PFC have been analyzed in food, in a way that it is not possible to establish the PFC homologue present in this matrix. Comprehensive food surveys and studies on gastrointestinal uptake are urgently required for a better understanding of the contribution of food pathway to consumer exposure to PFC. There is also a well-established record that should be highlighted: the ubiquitous presence and levels of PFOS and PFOA in human milk. These levels justify further monitoring of this class of contaminants worldwide. In any case, the understanding of exposures to PFC through the diet is still in its early phase, and only relatively few food samples have been analyzed in several countries. Further studies on the correlation between food intake and exposure, as well as food measurements, are needed before reliable conclusions can be made on the source of dietary exposures in humans.

On the assessment of dietary exposure to PFC, it is worth noting again that there are some limitations. First, a conservative estimation of PFC's dietary exposure is used because the analyzed food represents only a portion of the average diet and the dietary habits for the different groups of populations are not considered. Second, most studies do not consider precursor compounds that could be taken up and converted to PFOS or PFOA within the human body. Factors contributing to these limitations will be addressed in future studies.

References

1. Boulanger B, Vargo JD, Schnoor JL, Hornbuckle KC (2005) Environ Sci Technol 39: 5524–5530
2. Clara M, Scheffknecht C, Scharf S, Weiss S, Gans O (2008) Water Sci Technol 58(1):59–66
3. Nakano Y, Kimoto H, Watanabe S, Harada K, Watanabe YW (2006) J Oceanogr 62:71–81
4. Carmosini N, Lee LS (2008) Environ Sci Technol 42:6559–6565
5. Haukas M, Berger U, Hop H, Gulliksen B, Gabrielsen GW (2007) Environ Pollut 148: 360–371
6. Martin JW, Whittle DM, Muir DCG, Mabury SA (2004) Environ Sci Technol 38:5379–5385
7. Taniyasu S, Kannan K, Man KS, Gulkowska A, Sinclair E, Okazawa T, Yamashita N (2005) J Chromatogr A 1093:89–97
8. Holmström KE, Johansson AK, Bignert A, Lindberg P, Berger U (2010) Environ Sci Technol 44:4083–4088
9. D'Hollander W, De Voogt P, De Coen W, Bervoets L (2010) Rev Environ Contamin Toxicol 208:179–215

10. Butt CM, Berger U, Bossi R, Tomy GT (2010) Sci Total Environ 408:2936–2965
11. Kunacheva C, Fujii S, Tanaka S, Boontanon SK, Poothong S, Wongwatthana T, Shivakoti BR (2010) J Water Suppl Res Technol AQUA 59:345–354
12. Kannan K, Corsolini S, Falandysz J, Fillmann G, Kumar KS, Loganathan BG, Mohd MA, Olivero J, Van Wouwe N, Yang JH, Aldous KM (2004) Environ Sci Technol 38:4489–4495
13. Kärrman A, Harada KH, Inoue K, Takasuga T, Ohi E, Koizumi A (2009) Environ Int 35: 712–717
14. Kärrman A, van Bavel B, Järnberg U, Hardell L, Lindström G (2006) Chemosphere 64: 1582–1591
15. Kärrman A, Domingo JL, Llebaria X, Nadal M, Bigas E, van Bavel B, Lindström G (2010) Environ Sci Pollut Res 17:750–758
16. Llorca M, Farré M, Picó Y, Teijón ML, Álvarez JG, Barceló D (2010) Environ Int 36: 584–592
17. Bjärklund JA, Thuresson K, De Wit CA (2009) Environ Sci Technol 43:2276–2281
18. Ericson I, Martí-Cid R, Nadal M, Van Bavel B, Lindström G, Domingo JL (2008) J Agr Food Chem 56:1787–1794
19. Fromme H, Albrecht M, Angerer J, Drexler H, Gruber L, Schlummer M, Parlar H, Kärner W, Wanner A, Heitmann D, Roscher E, Bolte G (2007) Int J Hyg Environ Health 210:345–349
20. Tomy GT, Budakowski W, Halldorson T, Helm PA, Stern GA, Friesen K, Pepper K, Tittlemier SA, Fisk AT (2004) Environ Sci Technol 38:6475–6481
21. Kwadijk CJAF, Korytar P, Koelmans AA (2010) Environ Sci Technol 44:3746–3751
22. Kelly BC, Ikonomou MG, Blair JD, Surridge B, Hoover D, Grace R, Gobas FAPC (2009) Environ Sci Technol 43:4037–4043
23. Jeon J, Kannan K, Lim HK, Moon HB, Ra JS, Kim SD (2010) Environ Sci Technol 44: 2695–2701
24. Bhhatarai B, Gramatica P (2011) Molecular Diversity 15:467–476
25. Tichý M, Valigurová R, Čabala R, Uzlová R, Rucki M (2010) Interdiscip Toxicol 3:73–75
26. Phillips MM, Dinglasan-Panlilio MJA, Mabury SA, Solomon KR, Sibley PK (2010) Environ Toxicol Chem 29:1123–1131
27. Lau C, Butenhoff JL, Rogers JM (2004) Toxicol Appl Pharmacol 198:231–241
28. Fromme H, Schlummer M, Möller A, Gruber L, Wolz G, Ungewiss J, Böhmer S, Dekant W, Mayer R, Liebl B, Twardella D (2007) Environ Sci Technol 41:7928–7933
29. Rosen MB, Lee JS, Ren H, Vallanat B, Liu J, Waalkes MP, Abbott BD, Lau C, Corton JC (2008) Toxicol Sci 103:46–56
30. Ren H, Vallanat B, Nelson DM, Yeung LWY, Guruge KS, Lam PKS, Lehman-McKeeman LD, Corton JC (2009) Reproduct Toxicol 27:266–277
31. Voogt PD, Saez M (2006) Trend Anal Chem 25:326–342
32. Villagrasa M, Lopez De, Alda M, Barcelo D (2006) Anal Bioanal Chem 386:953–972
33. Houde M, Martin JW, Letcher RJ, Solomon KR, Muir DCG (2006) Environ Sci Technol 40: 3463–3473
34. Suja F, Pramanik BK, Zain SM (2009) Water Sci Technol 60:1533–1554
35. Giesy JP, Naile JE, Khim JS, Jones PD, Newsted JL (2010) Rev Environ Contamin Toxicol 202:1–52
36. Gulkowska A, Jiang Q, So MK, Taniyasu S, Lam PKS, Yamashita N (2006) Environ Sci Technol 40:3736–3741
37. Tseng CL, Liu LL, Chen CM, Ding WH (2006) J Chromatogr A 1105:119–126
38. Kannan K, Tao L, Sinclair E, Pastva SD, Jude DJ, Giesy JP (2005) Archives of Environmental Contamination and Toxicology 48:559–566
39. Bailey VA, Clarke D, Routledge A (2010) J Fluor Chem 131:691–697
40. Schuetze A, Heberer T, Effkemann S, Juergensen S Chemosphere 78:647–652
41. Ballesteros-Gómez A, Rubio S, van Leeuwen S (2010) J Chromatogr A 1217:5913–5921
42. Kuklenyik Z, Reich JA, Tully JS, Needham LL, Calafat AM (2004) Environ Sci Technol 38: 3698–3704

43. Berger U, Glynn A, Holmström KE, Berglund M, Ankarberg EH, Törnkvist A (2009) Chemosphere 76:799–804
44. Liu J, Li J, Zhao Y, Wang Y, Zhang L, Wu Y (2010) Environ Int 36:433–438
45. Del Gobbo L, Tittlemier S, Diamond M, Pepper K, Tague B, Yeudall F, Vanderlinden L (2008) J Agric Food Chem 56:7551–7559
46. Völkel W, Genzel-Boroviczny O, Demmelmair H, Gebauer C, Koletzko B, Twardella D, Raab U, Fromme H (2008) Int J Hyg Environ Health 211:440–446
47. Haug LS, Salihovic S, Jogsten IE, Thomsen C, van Bavel B, Lindström G, Becher G (2010) Chemosphere 80:1137–1143
48. Tittlemier SA, Pepper K, Seymour C, Moisey J, Bronson R, Cao XL, Dabeka RW (2007) J Agr Food Chem 55:3203–3210
49. Furdui VI, Stock NL, Ellis DA, Butt CM, Whittle DM, Crozier PW, Reiner EJ, Muir DCG, Mabury SA (2007) Environ Sci Technol 41:1554–1559
50. Sinclair E, Mayack DT, Roblee K, Yamashita N, Kannan K (2006) Arch Environ Contamin Toxicol 50:398–410
51. Skutlarek D, Exner M, Farber H (2006) Environ Sci Pollut Res 13:299–307
52. Martin JW, Kannan K, Berger U, De Voogt P, Field J, Franklin J, Giesy JP, Harner T, Muir DCG, Scott B, Kaiser M, Jarnberg U, Jones KC, Mabury SA, Schroeder H, Simcik M, Sottani C, Van Bavel B, Karrman A, Lindstrom G, Van Leeuwen S (2004) Environ Sci Technol 38:248A–255A
53. Hansen KJ, Clemen LA, Ellefson ME, Johnson HO (2001) Environ Sci Technol 35:766–770
54. Holm A, Wilson SR, Molander P, Lundanes E, Greibrokk T (2004) J Sep Sci 27:1071–1079
55. Loveless SE, Finlay C, Everds NE, Frame SR, Gillies PJ, O'Connor JC, Powley CR, Kennedy GL (2006) Toxicology 220:203–217
56. Liu J, Lee LS (2005) Environ Sci Technol 39:7535–7540
57. Fromme H, Tittlemier SA, Völkel W, Wilhelm M, Twardella D (2009) Int J Hyg Environ Health 212:239–270
58. Dinglasan MJA, Ye Y, Edwards EA, Mabury SA (2004) Environ Sci Technol 38:2857–2864
59. Szostek B, Prickett KB, Buck RC (2006) Rapid Commun Mass Spectr 20:2837–2844
60. Wang N, Szostek B, Folsom PW, Sulecki LM, Capka V, Buck RC, Berti WR, Gannon JT (2005) Environ Sci Technol 39:531–538
61. Schultz MM, Barofsky DF, Field JA (2006) Environ Sci Technol 40:289–295
62. Ylinen M, Hahijarvi H, Peura P, Ramo O (1985) Arch Environ Contamin Toxicol 14:713–717
63. Giesy JP, Kannan K (2001) Environ Sci Technol 35:1339–1342
64. Van De Vijver KI, Hoff PT, Van Dongen W, Esmans EL, Blust R, De Coen WM (2003) Environ Toxicol Chem 22:2037–2041
65. Taniyasu S, Kannan K, Horii Y, Hanari N, Yamashita N (2003) Environ Sci Technol 37:2634–2639
66. Berger U, Haukas M (2005) J Chromatogr A 1081:210–217
67. Guruge KS, Manage PM, Yamanaka N, Miyazaki S, Taniyasu S, Yamashita N (2008) Chemosphere 73:S210–S215
68. Tittlemier SA, Pepper K, Edwards L, Tomy G (2005) J Chromatogr A 1066:189–195
69. Fromme H, Tittlemier SA, Völkel W, Wilhelm M, Twardella D (2008) Int J Hyg Environ Health
70. Jogsten IE, Perelló G, Llebaria X, Bigas E, Martí-Cid R, Kärrman A, Domingo JL (2009) Food Chem Toxicol 47:1577–1583
71. Llorca M, Farré M, Picó Y, Barceló D (2009) J Chromatogr A 1216:7195–7204
72. Takino M, Daishima S, Nakahara T (2003) Rapid Commun Mass Spectrom 17:383–390
73. Chu S, Letcher RJ (2008) J Chromatogr A 1215:92–99
74. Song L, Wellman AD, Yao H, Adcock J (2007) Rapid Commun Mass Spectrom 21: 1343–1351
75. Llorca M, Farré M, Picó Y, Barceló D (2010) Anal Bioanal Chem:1–15

76. Ericson I, Domingo JL, Nadal M, Bigas E, Llebaria X, Van Bavel B, Lindström G (2009) Arch Environ Contamin Toxicol 57:631–638
77. Harada K, Saito N, Sasaki K, Inoue K, Koizumi A (2003) Bull Environ Contamin Toxicol 71:31–36
78. Wilhelm M, Bergmann S, Dieter HH (2010) Int J Hyg Environ Health 213:224–232
79. Loos R, Wollgast J, Huber T, Hanke G (2007) Anal Bioanal Chem 387:1469–1478
80. Martin JW, Mabury SA, Solomon KR, Muir DCG (2003) Environ Toxicol Chemi 22: 196–204
81. Ostertag SK, Tague BA, Humphries MM, Tittlemier SA, Chan HM (2009) Chemosphere 75:1165–1172
82. Haug LS, Thomsen C, Brantsaeter AL, Kvalem HE, Haugen M, Becher G, Alexander J, Meltzer HM, Knutsen HK (2010) Environ Int 36:772–778
83. Powley CR, George SW, Ryan TW, Buck RC (2005) Anal Chem 77:6353–6358
84. Begley TH, White K, Honigfort P, Twaroski ML, Neches R, Walker RA (2005) Food Addit Contam 22:1023–1031
85. Martin JW, Mabury SA, Solomon KR, Muir DCG (2003) Environ Toxicol Chem 22:189–195
86. de Vos MG, Huijbregts MAJ, van den Heuvel-Greve MJ, Vethaak AD, Van de Vijver KI, Leonards PEG, van Leeuwen SPJ, de Voogt P, Hendriks AJ (2008) Chemosphere 70:1766–1773
87. Gruber L, Schlummer M, Ungewiss J, Wolz G, Moeller A, Weise N, Sengl M, Frey S, Gerst M, Schwaiger J (2007) Organohalogen Compd 69:3
88. Ye X, Strynar MJ, Nakayama SF, Varns J, Helfant L, Lazorchak J, Lindstrom AB (2008) Environ Pollut 156:1227–1232
89. Tomy GT, Tittlemier SA, Palace VP, Budakowski WR, Braekevelt E, Brinkworth L, Friesen K (2004) Environ Sci Technol 38:758–762
90. Nania V, Pellegrini GE, Fabrizi L, Sesta G, Sanctis PD, Lucchetti D, Pasquale MD, Coni E (2009) Food Chem 115:951–957
91. Kärrman A, Ericson I, VanBavel B, Ola Darnerud P, Aune M, Glynn A, Ligneli S, Lindström G (2007) Environ Health Perspect 115:226–230
92. Van Den Heuvel-Greve M, Leonards P, Vethaak D (2006) Dioxin Research Westerschelde: Measuring Percentages of Dioxins80
93. Nakata H, Kannan K, Nasu T, Cho HS, Sinclair E, Takemura A (2006) Environ Sci Technol 40:4916–4921
94. Cunha I, Hoff P, Van De Vijver K, Guilhermino L, Esmans E, De Coen W (2005) Mar Pollut Bull 50:1128–1132
95. Tao L, Kannan K, Wong CM, Arcaro KF, Butenhoff JL (2008) Environ Sci Technol 42:3096–3101
96. De Voogt P, Berger U, De Coen W, De Wolf W, Heimstad E, McLachlan M, Van Leeuwen S, Van Roon A (2006) Perfluor Org Compds Eur Environ: http://www.science.uva.nl/perforce/Final%20reportA.pdf
97. Fsa U (2006) Fluorinated Chemicals: UK Dietary Intakes
98. Tittlemier SA, Pepper K, Edwards L (2006) J Agricult Food Chem 54:8385–8389
99. Loewen M, Halldorson T, Wang F, Tomy G (2005) Environ Sci Technol 39:2944–2951
100. US EPA (Environmental Protection Agency) AoP, FOSA, and PFOA from various food matrices using HPLC electrospray/mass spectrometry, 3 M study conducted by Centre Analytical Laboratories, Inc (http://www.ewg.org/files/), 200
101. Schecter A, Colacino J, Haffner D, Patel K, Opel M, Päpke O, Birnbaum L (2010) Environ Health Perspect 118:796–802
102. Lau C, Anitole K, Hodes C, Lai D, Pfahles-Hutchens A, Seed J (2007) Toxicol Sci 99:366–394
103. Ropers MH, Durand S, Veyrand B, Beaumal V, Marchand P, Anton M, Le Bizec B (2009) Food Hydrocolloid 23:1149–1155

104. Calafat AM, Ye X, Silva MJ, Kuklenyik Z, Needham LL, Kolossa M, Tuomisto J, Astrup Jensen A (2006) Int J Androl 29:166–171
105. Takagi S, Adachi F, Miyano K, Koizumi Y, Tanaka H, Mimura M, Watanabe I, Tanabe S, Kannan K (2008) Chemosphere 72:1409–1412
106. Schriks M, Heringa MB, van der Kooi MME, de Voogt P, van Wezel AP (2010) Water Res 44:461–476
107. Brede E, Wilhelm M, Gäen T, Möller J, Rauchfuss K, Kraft M, Hölzer J (2010) Int J Hyg Environ Health 213:217–223
108. Trudel D, Horowitz L, Wormuth M, Scheringer M, Cousins IT, HungerbÃ¼hler K (2008) Risk Anal 28:251–269
109. Hart K, Kannan K, Tao L, Takahashi S, Tanabe S (2008) Sci Total Environ 403:215–221
110. Phillips MM, Dinglasan-Panlilio MJA, Mabury SA, Solomon KR, Sibley PK (2007) Environ Sci Technol 41:7159–7163
111. Tao L, Ma J, Kunisue T, Libelo EL, Tanabe S, Kannan K (2008) Environ Sci Technol 42:8597–8602
112. Wang Y, Zhang P (2010) Prog Chem 22:210–219

Human Biomonitoring of Perfluorinated Compounds

Michael Wilhelm and Jürgen Hölzer

Abstract Human biomonitoring (HBM) is a scientific technique for assessing human exposures to natural and synthetic compounds in the environment. It is based on analysis of human tissues and fluids. It provides the only direct method of determining if people have been exposed to particular substances, what the magnitudes of their exposures are, and how these may be changing over time. In HBM, the most commonly studied perfluorinated compounds are the perfluorinated sulfonates and the perfluorinated carboxylates. Among these perfluorooctane sulfonate (PFOS) and perfluorooctanoate (PFOA) are of greatest concern. Our first biomonitoring study from autumn 2006 evidenced that plasma PFOA concentrations of residents from Arnsberg were 4.5–8.3 times higher than those in reference groups. A 10–20% reduction of PFOA plasma levels in residents from Arnsberg, Sauerland, Germany, was observed in our 1-year follow-up study. A further but still slow decline of the PFOA load was confirmed in the 2-year follow-up study. Detailed monitoring of perfluorinated compounds in the Region Sauerland also revealed high contamination of fish with PFOS. We observed high PFOS levels in plasma of anglers which was clearly related to the consumption of fish caught from the Möhnelake.

Due to uncertainties and inconsistencies in the epidemiological studies, no health-based HBM values for perfluorinated compounds in blood could be set from the available data yet. A further approach to interpret perfluorinated compounds levels in HBM is to derive HBM values from corresponding tolerable intake doses, such like the tolerable daily intake (TDI). This concept has been proposed by the German Human Biomonitoring Commission (2007). The Commission is aware of the uncertainties of such derivation and estimates.

M. Wilhelm (✉) • J. Hölzer
Department of Hygiene, Social and Environmental Medicine, Ruhr-University Bochum, Universitätsstraße 150, 44801 Bochum, Germany
e-mail: wilhelm@hygiene.rub.de

T.P. Knepper and F.T. Lange (eds.), *Polyfluorinated Chemicals and Transformation Products*, Hdb Env Chem (2012) 17: 155–168,
DOI 10.1007/978-3-642-21872-9_8,

Keywords Concentrations • Exposure • Human biomonitoring (HBM) • Perfluorinated compounds • Toxicity

Contents

Abbreviations

APFO	Ammonium perfluorooctanoate
HBM	Human biomonitoring
PFC	Perfluorinated compounds
PFOS	Perfluorooctane sulfonate
PFOA	Perfluorooctanoate
PFHxS	Perfluorohexane sulfonate
LOQ	Limit of quantification
HPLC	High performance liquid chromatography
HDL	High-density lipoprotein
LDL	Low-density lipoprotein
TDI	Tolerable daily intake

1 General Aspects on Human Biomonitoring

Human biomonitoring (HBM) is a scientific technique for assessing human exposures to natural and synthetic compounds in the environment. It is based on analysis of human tissues and fluids. It provides the only direct method of determining if people have been exposed to particular substances, what the magnitudes of their exposures are, and how these may be changing over time. HBM has become a more useful tool in recent years as the result of advancements in the capability to measure more and more minute amounts of chemicals in the human body. HBM considers all routes of uptake and all sources which are relevant making it an ideal instrument for risk assessment and risk management. HBM can identify new chemical exposures, trends and changes in exposure, establish distribution of exposure among the general

Table 1 List of perfluorinated compounds as measured by Calafat et al. [2] and Haug et al. [3] in serum

Perfluorinated compounds	Abbreviation	CAS number Free acids
Perfluorobutanoate	PFBA	375-22-4
Perfluoropentanoate	PFPeA	2,706-90-3
Perfluorohexanoate	PFHxA	307-24-4
Perfluoroheptanoate	PFHpA	375-85-9
Perfluoroctanoate	PFOA	335-67-1
Perfluorononanoate	PFNA	375-95-1
Perfluorodecanoate	PFDeA	335-76-2
Perfluoroundecanoate	PFUA	2,058-94-8
Perfluorododecanoate	PFDoA	307-55-1
Perfluorotridecanoate	PFTrDA	72,629-94-8
Perfluorotetradecanoate	PFTeDA	376-06-7
Perfluorobutane sulfonate	PFBS	375-73-5
Perfluorohexane sulfonate	PFHxS	355-46-4
Perfluoroheptane sulfonate	PFHpS	375-92-8
Perfluoroctane sulfonate	PFOS	1,763-23-1
Perfluorodecane sulfonate	PFDS	335-77-3
Perfluorooctane sulfonamide	PFOSA	754-91-6
2-(N-ethyl-perfluorooctane sulfonamido) acetic acid	Et-PFOSA-AcOH	
2-(N-methyl-perfluorooctane sulfonamido) acetic acid	Me-PFOSA-AcOH	

population, identify vulnerable groups and populations with higher exposures. Blood and urine are by far the most approved matrices. HBM can be done for most chemical substances which are in the focus of the worldwide discussion of environmental medicine. This especially applies for metals, polycyclic aromatic hydrocarbons, phthalates, dioxins, pesticides, aromatic amines, environmental tobacco smoke as well as for perfluorinated compounds (PFC). More details on HBM have been reviewed recently by Angerer et al. [1].

1.1 Perfluorinated Compounds Studied in HBM

In HBM, the most commonly studied PFC are the perfluorinated sulfonates and the perfluorinated carboxylates. Among these perfluorooctane sulfonate (PFOS) and perfluorooctanoate (PFOA) are of greatest concern. Both persist in humans and the environment. Besides PFOS and PFOA, perfluorohexane sulfonate (PFHxS) is also frequently detected in human samples. Among the National Health and Nutrition Examination Survey (NHANES) conducted by the U.S. Centers for Disease Control and Prevention 12 PFC are measured regularly in human serum [2]. Haug et al. [3] included 19 PFC in their study with serum samples from Norway residents (Table 1).

In NHANES, the levels of the following PFC are mostly (more than 50%) below limit of detection: PFBS, PFDeA, PFDoA, PFHpA, PFOSA, PFUA, Et-PFOSA-AcOH and Me-PFOSA-AcOH. Limits of detection in serum varies between 0.1 (PFOA, PFNA) and 1.0 μg/L (PFDoA) [2]. In Haug et al. [3] PFBA, PFHxA, PFTeDA, PFDS, Et-PFOSA-AcOH and Me-PFOSA-AcOH were not observed above limit of quantification (LOQ) (0.05–0.1 μg/L). PFOS, PFOA, PFHxS, PFNA, PFDA, PFUA, PFTrDA, PFHpS and PFOSA were detected in most samples.

2 Determination of PFC in Human Biomonitoring

Sensitive methods are available to measure PFC in serum, plasma and breast milk (e.g., [4, 5]). Usually, the PFC are extracted from interfering matrix compounds by solid phase extraction. After elution, the PFC are chromatographically separated by high performance liquid chromatography (HPLC) and detected by tandem mass spectrometry (MS/MS). Calibration is performed using standard solutions prepared in bovine serum which are treated in the same manner as the human plasma samples analyzed. As internal standards ^{13}C- or ^{18}O-labeled analogues of PFC are used. Limits of detection are in the range of 0.05–1.0 μg/L. For quality control participation in different quality assessment schemes is possible (e.g., German External Quality Assessment Scheme, G-EQUAS, Erlangen, Germany).

An interlaboratory study with six laboratories showed that the analysis of PFC in blood matrices can be done with good precision among people in background-exposed populations [6]. All laboratories used HPLC-MS/MS. The average within- and between-batch coefficient of variation for PFOS was 9.1% and 9.3%; for PFOA was 14.5% and 14.5%; and for PFHxS was 14.5% and 17.0%.

The internal exposure is estimated based on concentrations in plasma, serum or whole blood. Validation studies have shown that serum and plasma samples yield comparable results regarding PFOS, PFOA and PFHxS concentrations [7]. Is it assumed that levels in whole blood are 50% below levels in serum or plasma, although the results are not consistent. Samples with widely differing concentrations were analyzed by Ehresman et al. [7] and a median plasma to whole blood ratio of 2.3 was observed for PFOS (ranges: 1.8–3.3 and 1.8–2.9 for whole blood collected in EDTA and heparin, respectively). For PFOA, the median ratio was 2.0; for PFHxS ratios were 2.4 or 2.1 depending on the anticoagulant used. A contrasting result was published by Kärrman et al. [8], who analyzed whole blood and plasma samples from five subjects. They found a plasma to whole blood ratio of 1.2 (PFHxS), 1.4 (PFOA), 1.2 (PFOS), 1.0 (PFNA) and 0.2 for PFOSA. Most studies nowadays measure PFC in plasma or serum.

Some HBM studies on PFC levels in breast milk are also available (e.g., [9]). PFC concentrations in breast milk are much lower compared with those in plasma. However, PFC intake via breast milk leads to a body burden in infants at the age of 6 months similar to (PFOS) or higher than (PFOA) that found in adults [10].

3 PFC Concentrations in Blood Plasma/Serum

PFOA and PFOS have been detected globally in human blood samples. The highest PFOS and PFOA concentrations were measured in workers employed in fluorine production plants [11]. The exposure of the general population differs between countries. Fromme et al. [12] recently summarized HBM data and reported that mean concentrations for some PFC from North American populations appear to be slightly higher than European, Asian and Australian populations studied.

In less industrialized countries PFC exposure occurs on a very low level. Among a HBM study performed with children and adults living in Afghanistan PFOS could be quantified in all blood samples (LOQ: 0.1 μg/L), median value was 1.2 μg/L. Most PFOA and PFHxS levels were below LOQ of 0.5 μg/L [13]. Similar observations have been reported for India [14], Sri Lanka [15] and for Peruvian residents [16].

Increased PFOA exposure of general population groups mainly occurred via ingestion of contaminated drinking water: Mid-Ohio Valley, USA [17, 18] and Arnsberg, Sauerland area, Germany [19]. In Arnsberg, Germany, 40,000 residents had been exposed to PFOA-contaminated drinking water (500–640 ng/L PFOA; May 2006).

Median and maximum PFOS and PFOA concentrations reported from occupationally exposed workers, the C8 Health Project, German residents exposed to contaminated drinking water and the National Health and Nutrition Examination Survey (NHANES, USA) are compared in Fig. 1.

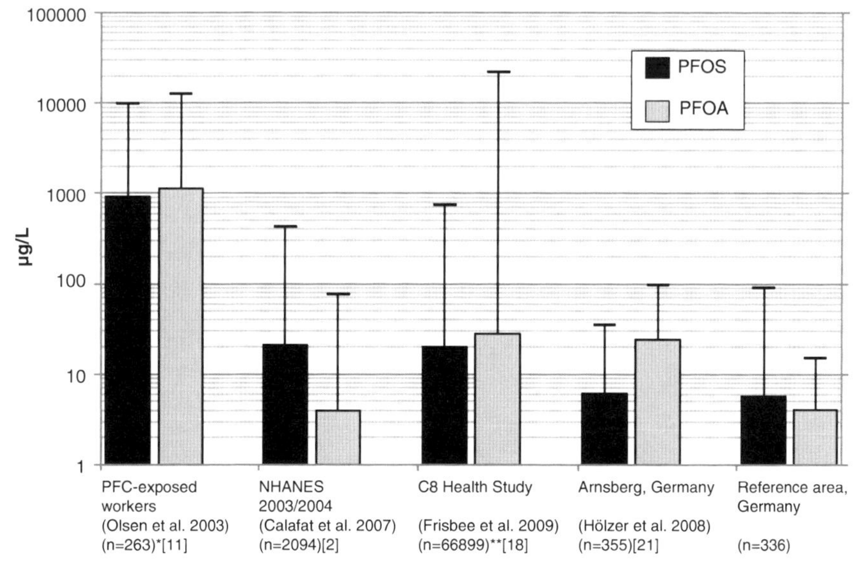

Fig. 1 PFOA and PFOS concentrations in blood plasma. Median (*bars*) and maximum (*whiskers*) concentrations cited from selected international human biomonitoring studies. * geometric mean, ** maxima from [20]

3.1 *Exposure of German Residents in Arnsberg to PFC-Contaminated Drinking Water*

During their investigations into PFC concentrations of surface waters in Germany, Skutlarek et al. [22] observed remarkably high PFOA concentrations not only in the rivers Ruhr (tributary of the Rhine, up to 177 ng/L) and Möhne (tributary of the Ruhr, up to 7,070 ng/L), but also in public water supplies, which use river water to produce drinking water by bank filtration or artificial recharge. The highest PFC concentration in drinking water, which was reported by Skutlarek et al., was 598 ng/L. Based on the results of an extensive environmental monitoring program, federal health authorities concluded that PFC contamination of agricultural land occurred by the widespread use of soil conditioner, which had been mingled with industrial waste. In July 2006, activated charcoal filters were installed that efficiently decreased PFOA concentrations in drinking water to levels predominantly under the LOQ (10 ng/L). In September and October 2006, 355 residents from Arnsberg who had been supplied by contaminated drinking water, together with 336 residents from neighboring towns Siegen and Brilon, who received water with PFOA levels below the LOQ, were included in a first biomonitoring study. In both locations, school beginners and their mothers were asked to participate. Geometric mean levels of PFOA plasma concentration of children, women and men from Arnsberg were 22.1 μg/L, 23.4 μg/L and 25.3 μg/L, respectively. They were increased 4.5–8.3-fold in comparison to PFOA levels in the control population. Consumption of PFC-contaminated tap water was a significant predictor of PFOA plasma concentrations [21]. The study group has been followed up, last study was 2010. Yearly follow-up studies show a decline of PFOA plasma levels, after charcoal filtration was introduced in July 2006 at the water works [23, 24].

3.2 *Exposure of U.S. American Residents in the Mid-Ohio Valley to PFC: C8 Health Project*

The world's largest HBM study is the C8 Health Project. It was created, authorized and funded as part of the settlement agreement reached in the case of Jack W. Leach, et al. v. E.I. du Pont de Nemours & Company. The reason was the perfluorooctanoic acid (PFOA, or C8) contamination of drinking water in six water districts in two states near the DuPont Washington Works facility near Parkersburg, West Virginia. 69,030 residents took part (enrollment was 2005–2006). Extensive data were collected, including demographic data, medical diagnoses, clinical laboratory testing and determination of serum concentrations of 10 PFC were performed. Results are reported on the C8 Health Project homepage (http://www.hsc.wvu.edu/som/cmed/c8/index.asp) and on the pages of the C8 Science Panel (http://www.c8sciencepanel.org/index.html). The population geometric mean for serum PFOA was 32.9 μg/L. Serum concentrations for PFHxS and PFNA were elevated 39 and 73%, respectively, whereas PFOS was present at levels similar to those in the U.S. population [18].

4 Elimination Half-Life

PFC half-life of elimination from blood differs between species (much shorter in animals compared to humans) and between PFC (the longer the chain length, the longer the half-life) (overview in [25]). Half-lives in cynomolgus monkeys for PFBS (3.5–4 days) and PFBA (1.7 days) are much shorter than those for PFOS (150 days) and PFOA (30 days females, 21 days males) [26]. Elimination half-lives of a series of PFC were studied in male and female cynomolgus monkeys following intravenous injections [27]. Half-lives were as follows:

PFBA (1.68 days males; 1.71 females) $\approx$ PFHxA (1.45 days males; 0.81 days females)<<<< PFOA (20.9 days males; 32.6 days females).

PFBS (4 days males; 3.5 days females)<<< PFHxS (141 days males; 87 days females) $\approx$ PFOS (132 days males; 110 days females).

Among six human subjects (5 male, 1 female) followed up to 180 days, Olsen et al. [28] found a geometric mean serum elimination half-life for PFBS of 25.8 days (95% confidence interval, 16.6–40.2). Chang et al. [27] studied 177 individuals with potential exposure to PFBA through drinking water. Mean terminal serum PFBA elimination half-life was 74.6 h. More data are available on PFOA and PFOS half-lives in humans. Olsen et al. [28] studied serum elimination half-life for a group 26 retired fluorochemical production workers. They estimated the following geometric mean half-lives: 4.8 (95% CI, 4.0–5.8) years for PFOS; 3.5 (95% CI, 3.1–4.4) years for PFOA. Within a follow-up study with residents from Ohio and West Virginia a median half-life for PFOA of 2.3 years (95% confidence interval, 2.1–2.4) was reported by Bartell et al. [29]. However, data on PFHxS are somewhat different. In a study with retired workers [28] PFHxS had a longer elimination half-life (7.3; 95% CI 5.8–9.2 years) compared to PFOS (4.8). In monkeys PFHxS half-life (87 days females, 141 days males) was similar to PFOS (150 days) [26].

Furthermore, there are indications that elimination rate slightly differs between the perfluoroalkylcarboxylic acids (e.g., PFBA, PFPA, PFHxA, PFHpA, PFOA) and the perfluoroalkylsulphonic acids (e.g., PFBS, PFPS, PFHxS, PFHpS, PFOS). Data to support this view are available for the C4 and C8 PFC. Half-lives for PFBS (3.5–4 days) and PFOS are longer than those for PFBA and PFOA, respectively [26–28]. The explanation is the presence of always one more fluorinated C-atom in the perfluorosulfonates as compared to the carboxylates.

5 Time Trends of PFC Levels in HBM

Several studies on time trends of the internal exposure are available. Some consistent conclusions can be drawn. In some industrialized countries PFC levels show increasing levels until around 1990–2000. Trend analyses indicate a reduction of the internal exposure particularly in the years following 2001. This apparent reduction in PFOS concentrations may be related to the cessation in production of perfluorooctylsulfonyl compounds that began in 2000.

In studies from the U.S., serum levels of PFOS and PFOA increased from 1974 to 1989 and since about 2000 a decrease of PFOA, PFOS and PFHxS serum concentrations has been reported [2, 11, 30–32].

In Japan, the analysis of serum samples collected 1983–1999 showed a significant increase in PFOA levels, while for PFOS no such increase could be observed [33]. Another Japanese study reported rising PFOS levels and to a greater amount increasing PFOA levels for the period of 1977–2003 in Miyagi, only a discrete PFOS increase could be measured in Akita [34].

Results of a Chinese study that analyzed serum samples from 1987 to 2002 also showed a considerable increase in PFOS and PFOA concentrations during this time period [35]. In Norwegian residents a ninefold increase in the serum concentrations of PFOS, PFOA and PFHpS was measured for men (age 40–50 years) from 1977 to the mid-1999s [3]. The concentrations reached then a plateau and started to decrease around the year 2000. A similar trend was also seen for PFNA, PFHxS, PFDoDA, PFUA.

The follow-up study of the Arnsberg PFC case [19, 21] showed a decrease of PFOS, PFOA and PFHxS plasma levels between 2006 and 2008 in the control group (German adults and children) [23]. The geometric means of PFOA plasma levels declined by 13–15%.

In human milk samples from Sweden sampled between 1996 and 2004, no trend of PFC levels was observed [36]. However, in a more recent study, the concentrations of PFOS, PFHxS and PFOA in pooled human milk samples obtained in Sweden between 1972 and 2008 showed significant increasing trends from 1972 to 2000, with concentrations reaching a plateau in the 1990s. PFOA and PFOS showed statistically significant decreasing trends during 2001–2008 [37]. At the end of the study, in 2008, the measured concentrations of PFOS, PFHxS and PFOA in pooled human milk were 0.075 μg/L, 0.014 μg/L and 0.074 μg/L, respectively.

6 How to Interpret PFC Levels in HBM?

In Germany, basic principles of HBM have been defined by the German Human Biomonitoring Commission (http://www.umweltbundesamt.de/gesundheit-e/monitor/index.htm). The Commission was established in 1992 at the Federal Environment Agency. Two kinds of guidance values in HBM have been developed: the reference

Table 2 Reference values for PFOA and PFOS in plasma of the German population [38, 39]

	Adult males	Adult females	Children <10 years
PFOS μg/L	25	20	10
PFOA μg/L	10	10	10

value and the health-based HBM values. The reference value for a chemical substance in human biological material (e.g., blood, urine) is derived according to a defined statistical method from a series of measuring results obtained. Samples to be used for this purpose have to be collected employing a defined group of the general population. According to IUPAC guideline the Human Biomonitoring Commission uses as reference value the 95th percentile of the measured pollutant concentration levels in the relevant matrix of the reference population. To derive it, it is rounded off within the 95% confidence interval. In addition, when the data base is appropriate and sufficient to do so, the Commission defines reference values for sub-groups being subject or not to specific exposures (e.g., non-smoking and cadmium in blood). Wherever possible, reference values are defined using data obtained for a suitable reference population, such as the population studied in the German Environmental Surveys (GerES).

For PFC no data were available from the representative GerES. Based on three HBM studies, reference values for PFOS and PFOA in plasma were set (Table 2).

A repeat measurement should be performed in cases of concentrations exceeding the reference value. If reliable measurements show a value above the reference value, they should induce an environmental medicine-based search for sources. Such search should be carried out in a proportionate way.

Dietary intake seems to be the most important path of exposure for the general population to PFOS and PFOA [12]. Contamination of drinking water led to significantly increased PFOA concentration in blood samples of the affected populations in Little Hocking, Ohio, USA (Emmet et al. 2006) and Arnsberg, Germany [21]. Fish is an important part of the diet and recently the significance of fish consumption on the internal exposure to PFOS was emphasized [40–42].

As mentioned before workers occupationally exposed to PFC may have much higher PFC levels in blood than the general population.

It must be emphasized that reference values are statistically derived and do not represent toxicologically derived biological exposure limits. Thus, they cannot be used for health-related evaluation of HBM data. Nevertheless, the reference values permit to assess the exposure of individuals or population groups compared to the ubiquitous background exposure.

Health-based HBM values can be derived on the basis of epidemiological studies, toxicological basis with toxicokinetic extrapolation which provides a concentration of a substance or its metabolites corresponding to tolerable intake doses. However, HBM values for PFC have not been derived yet.

From epidemiological studies, no clear conclusions on associations of PFC levels in blood and health outcome can be drawn.

Lundin et al. [43] conducted a mortality study in a cohort of 3,993 highly exposed employees of an ammonium perfluorooctanoate (APFO) manufacturing facility. APFO rapidly dissociates to PFOA in blood. PFOA in serum was not associated with liver, pancreatic and testicular cancer or cirrhosis of the liver. Exposure was associated (albeit inconsistently) with prostate cancer, cerebrovascular disease and diabetes.

Sakr et al. [44] conducted a longitudinal study on 454 highly exposed workers and found inconsistent relationships between serum PFOA and lipids and liver enzymes: increase in total cholesterol, no association with triglycerides or other lipoproteins, association with total bilirubin and serum aspartate aminotransferase, but not with the other liver enzymes.

A retrospective cohort mortality analysis of 6,027 workers at a West Virginia fluoropolymer manufacturing plant found no increased risks for liver, pancreatic and testicular cancer [45]. Diabetes mortality was increased.

Influence of PFOS/PFOA serum/plasma levels of pregnant women and in newborns on birth outcome has been studied in several investigations. Inverse relationship on birth weight was found by Fei et al. [46] and Apelberg et al. [47]. Other authors reported no associations [48, 49]. Washino et al. [50] observed a negative association for PFOS, but not for PFOA.

In contrast, markedly elevated PFOA exposure via drinking water (Little Hocking, Ohio), as categorized by water service category, was not found to be associated with increased risk of lowered birth weight or gestational age [51]. Fei et al. [52] found that PFOA and PFOS exposure at plasma levels seen in the general population may reduce fecundity.

Several studies among residents with PFOA water contamination from the Mid-Ohio Valley show modest associations with health outcome and clinical chemical parameters (http://www.c8sciencepanel.org).

Higher PFOA and PFOS serum levels in children were associated with higher total cholesterol and low-density lipoprotein (LDL). Higher PFOS was associated with higher high-density lipoprotein (HDL), but not with triglycerides. No consistent trend was observed for PFOA and HDL or triglycerides.

Among nearly 50,000 participants over age 18 higher PFC was linked to higher cholesterol and higher uric acid. No association to diabetes was found.

In the study on puberty outcome there was a relationship of reduced odds of reached puberty in boys with increasing PFOS (delay of 190 days between the highest and lowest quartiles); for girls, higher exposure both PFOA and PFOS was associated with reduced odds of post-menarche (130 and 138 days of delay, respectively). For more details see website of C8 panel study.

The main authors (Kyle Steenland, Tony Fletcher, David Savitz) emphasize that the results of the C8 study have to interpreted with caution due to the cross-sectional study design and lack of knowledge on the mechanisms by which PFC act. Furthermore, other yet not identified factors than PFOA or PFOS may cause the observed associations.

Due to the mentioned uncertainties and inconsistencies in the epidemiological studies, no health-based HBM values for PFC in blood could be set from these data yet. A further approach to interpret PFC levels in HBM is to derive HBM values from corresponding tolerable intake doses, such like the tolerable daily intake (TDI). This concept has been proposed by the German Human Biomonitoring Commission [53] and is similar to the Biomonitoring Equivalents published by Hays et al. [54]. The Commission is aware of the uncertainties of such derivation and estimates.

Hepatotoxicity, developmental toxicity, immunotoxicity, hormonal effects and also a weak carcinogenic potential in animal studies have been described as main endpoints of health concern (summary in [26]). After detection of PFOA in drinking water at up to 640 ng/L in the city of Arnsberg, Germany, by Skutlarek et al. [22], Germany's Drinking Water Commission assessed PFC in drinking water and set in June 2006 a health-based guide value for safe lifelong exposure at 300 ng/L (sum of PFOA and PFOS) (summary in [25]) based on a TDI of 0.1 $\mu g/kg_{bw}$ and day for PFOA. The TDI was calculated from an estimated NOAEL of 0.1 mg/kg_{bw} for reproductive toxicity of PFOA in rats. For extrapolation on humans two extrapolation factors (EF) of ten each for inter- and intraspecies biologic variability and an additional safety factor (SF) of ten to cope with uncertainties due the much longer elimination half-life of PFOA in humans than in rats were applied ($TDI = 0.1\ (mg/kg_{bw})/1{,}000 = 0.1\ \mu g/kg_{bw}$) [25]. The same TDI is also applied for PFOS. It should be noted that the CONTAM Panel of EFSA [European Food Safety Authority (EFSA)] has derived other TDIs in 2008: for PFOA 1.5 $\mu g/kg_{bw}$ and for PFOS 0.15 $\mu g/kg_{bw}$ (http://www.efsa.europa.eu/en/efsajournal/pub/653.htm). However, HBM values according to this concept could not be set finally. At present there is an ongoing discussion on the toxicokinetic modeling. There are uncertainties on resorption, metabolism, elimination rate and intake-excretion ratio, including intra- and interindividual differences (such as age and gender).

Due to increasing knowledge on toxicokinetics and health-associated data from epidemiological studies, it seems feasible to derive HBM values for PFOS and PFOA in the near future.

References

1. Angerer J, Ewers U, Wilhelm M (2007) Human biomonitoring: state of the art. Int J Hyg Environ Health 210:201–228
2. Calafat AM, Wong LY, Kuklenyik Z, Reidy JA, Needham LL (2007) Polyfluoroalkyl chemicals in the U.S. population: data from the National Health and Nutrition Examination Survey (NHANES) 2003–2004 and comparisons with NHANES 1999–2000. Environ Health Perspect 115(11):1596–602
3. Haug LS, Thomsen C, Becher G (2009) Time trends and the influence of age and gender on serum concentrations of perfluorinated compounds in archived human samples. Environ Sci Technol 43:2131–2136
4. Haug LS, Thomsen C, Becher G (2009) A sensitive method for determination of a broad range of perfluorinated compounds in serum suitable for large-scale human biomonitoring. J Chromatogr A 1216:385–393

5. Mosch C, Kiranoglu M, Fromme H, Völkel W (2010) Simultaneous quantification of perflouroalkyl acids in human serum and breast milk using on-line sample preparation by HPLC column switching coupled to ESI-MS/M. J Chromatogr B Anal Technol Biomed Life Sci 878:2652–2658
6. Longnecker MP, Smith CS, Kissling GE, Hoppin JA, Butenhoff JL, Decker E, Ehresman DJ, Ellefson ME, Flaherty J, Gardner MS, Langlois E, Leblanc A, Lindstrom AB, Reagen WK, Strynar MJ, Studabaker WB (2008) An interlaboratory study of perfluorinated alkyl compound levels in human plasma. Environ Res 107:152–9
7. Ehresman DJ, Froehlich JW, Olsen GW, Chang S-C, Butenhoff JL (2007) Comparison of human whole blood, plasma, and serum matrices for the determination of perfluorooctanesulfonate (PFOA), and other fluorochemicals. Environ Res 103:176–184
8. Kärrman A, van Bavel B, Järnberg U, Hardell L, Lindström G (2006) Perfluorinated chemicals in relation to other persistent organic pollutants in human blood. Chemosphere 64:1582–1591
9. Völkel W, Genzel-Boroviczény O, Demmelmair H, Gebauer C, Koletzko B, Twardella D, Raab U, Fromme H (2008) Perfluorooctane sulphonate (PFOS) and perfluorooctanoic acid (PFOA) in human breast milk: results of a pilot study. Int J Hyg Environ Health 211:440–446
10. Fromme H, Mosch C, Morovitz M, Alba-Alejandre I, Boehmer S, Kiranoglu M, Faber F, Hannibal I, Genzel-Boroviczény O, Koletzko B, Völkel W (2010) Pre- and postnatal exposure to perfluorinated compounds (PFCs). Environ Sci Technol 44:7123–7129
11. Olsen GW, Church TR, Miller JP, Burris JM, Hansen KJ, Lundberg JK, Armitage JB, Herron RM, Medhdizadehkashi Z, Nobiletti JB, O'Neill EM, Mandel JH, Zobel LR (2003) Perfluorooctane sulfonate and other fluorochemicals in the serum of American Red Cross adult blood donors. Environ Health Perspect 111:1892–1901
12. Fromme H, Tittlemier SA, Völkel W, Wilhelm M, Twardella D (2009) Perfluorinated compounds – exposure assessment for the general population in Western countries. Int J Hyg Environ Health 212:239–270
13. Hemat H, Wilhelm M, Völkel W, Mosch C, Fromme H, Wittsiepe J (2010) Low serum levels of perfluorooctanoic acid (PFOA), perfluorooctane sulfonate (PFOS) and perfluorohexane sulfonate (PFHxS) in children and adults from Afghanistan. Sci Total Environ 408:3493–3495
14. Kannan K, Corsolini S, Falandysz J, Fillmann K, Kumar KS, Loganathan BG, Mohd MA, Olivero J, Van Wouwe N, Yang JH, Aldous KM (2004) Perfluorooctanesulfonate and related fluorochemicals in human blood from several countries. Environ Sci Technol 38:4489–4495
15. Guruge KS, Taniyasu S, Yamashita N, Wijeratna S, Mohotti KM, Seneviratne HR, Kannan K, Yamanaka N, Miyazaki S (2005) Perfluorinated organic compounds in human serum and seminal plasma: a study of urban and rural tea worker populations in Sri Lanka. J Environ Monit 7:371–377
16. Calafat AM, Needham LL, Kuklenyik Z, Reidy JA, Tully JS, Aguilar-Villalobos M, Naeher LP (2006) Perfluorinated chemicals in selected residents of the American continent. Chemosphere 63:490–496
17. Emmett EA, Shofer FS, Zhang H, Freeman D, Desai C, Shaw LM (2006) Community exposure to perfluorooctanoate: relationships between serum concentrations and exposure sources. J Occup Environ Med 2006(48):759–770
18. Frisbee SJ, Brooks AP Jr, Maher A, Flensborg P, Arnold S, Fletcher T, Steenland K, Shankar A, Knox SS, Pollard C, Halverson JA, Vieira VM, Jin C, Leyden KM, Ducatman AM (2009) The C8 health project: design, methods, and participants. Environ Health Perspect 117:1873–1882
19. Wilhelm M, Kraft M, Rauchfuss K, Hölzer J (2008) Assessment and management of the first German case of a contamination with perfluorinated compounds (PFC) in the Region Sauerland, North Rhine-Westphalia. J Toxicol Environ Health A 71:725–733
20. C8 Health Project (2008) Population Results for C8 and C8 Sulfonate Plasma Concentrations. http://www.hsc.wvu.edu/som/cmed/c8/results/C8AndC8SulfonateLevels/pdfs/Overall%20C8%20&%20C8S%20Results.pdf. Accessed 1st February 2011

21. Hölzer J, Midasch O, Rauchfuss K, Kraft M, Reupert R, Angerer J, Kleeschulte P, Marschall N, Wilhelm M (2008) Biomonitoring of perfluorinated compounds in children and adults exposed to perfluorooctanoate-contaminated drinking water. Environ Health Perspect 116:651–657
22. Skutlarek D, Exner M, Färber H (2006) Perfluorinated surfactants in surface and drinking waters. Environ Sci Pollut Res Int 13:299–307
23. Brede E, Wilhelm M, Göen T, Müller J, Rauchfuss K, Kraft M, Hölzer J (2010) Two-year follow-up biomonitoring pilot study of residents' and controls' PFC plasma levels after PFOA reduction in public water system in Arnsberg, Germany. Int J Hyg Environ Health 213:217–223
24. Hölzer J, Göen T, Rauchfuss K, Kraft M, Angerer J, Kleeschulte P, Wilhelm M (2009) One-year follow-up of perfluorinated compounds in plasma of German residents from Arnsberg formerly exposed to PFOA-contaminated drinking water. Int J Hyg Environ Health 212:499–504
25. Wilhelm M, Bergmann S, Dieter HH (2010) Occurrence of perfluorinated compounds (PFCs) in drinking water of North Rhine-Westphalia, Germany and new approach to assess drinking water contamination by shorter-chained C4–C7 PFCs. Int J Hyg Environ Health 213:224–232
26. Lau C, Anitole K, Hodes C, Lai D, Pfahles-Hutchens A, Seed J (2007) Perfluoroalkyl acids: a review of monitoring and toxicological findings. Toxicol Sci 99:366–394
27. Chang SC, Das K, Ehresman DJ, Ellefson ME, Gorman GS, Hart JA, Noker PE, Tan YM, Lieder PH, Lau C, Olsen GW, Butenhoff JL (2008) Comparative pharmacokinetics of perfluorobutyrate in rats, mice, monkeys, and humans and relevance to human exposure via drinking water. Toxicol Sci 104:40–53
28. Olsen GW, Burris JM, Ehresmann DJ, Froehlich JW, Seacat AM, Butenhoff JL, Zobel LR (2008) Half-life of serum elimination of perfluorooctanesulfonate, perfluorohexanesulfonate, and perfluorooctanesulfonate in retired fluorochemical production workers. Environ Health Perspect 115:1298–1305
29. Bartell SM, Calafat AM, Lyu C, Kato K, Ryan PB, Steenland K (2010) Rate of decline in serum PFOA concentrations after granular activated carbon filtration at two public water systems in Ohio and West Virginia. Environ Health Perspect 118:222–228
30. Olsen GW, Huang H-Y, Helzlsouer KJ, Hansen KJ, Butenhoff JL, Mandel JH (2005) Historical comparison of perfluorooctanesulfonate, perfluorooctanoate, and other fluorochemicals in human blood. Environ Health Perspect 113:539–545
31. Olsen GW, Mair DC, Reagan WK, Ellefson ME, Ehresman DJ, Butenhoff JL, Zobel LR (2007) Preliminary evidence of a decline in perfluorooctanesulfonate (PFOS) and perfluorooctanoate (PFOA) concentrations in American Red Cross blood donors. Chemosphere 68:105–111
32. Olsen GW, Mair DC, Church TR, Ellefson ME, Reagen WK, Boyd TM, Herron RM, Medhdizadehkashi Z, Nobiletti JB, Rios JA, Butenhoff JL, Zobel LR (2008) Decline in perfluorooctanesulfonate and other polyfluoroalkyl chemicals in American Red Cross adult blood donors, 2000–2006. Environ Sci Technol 42:4989–4995
33. Harada K, Koizumi A, Saito N, Inoue K, Yoshinaga T, Date C, Fujii S, Hachiya N, Hirosawa I, Koda S, Kusaka Y, Murata K, Omae K, Shimbo S, Takenaka K, Takeshita T, Todoriki H, Wada Y, Watanabe T, Ikeda M (2007) Historical and geographical aspects of the increasing perfluorooctanoate and perfluorooctane sulfonate contamination in human serum in Japan. Chemosphere 66:293–301
34. Harada K, Saito N, Inoue K, Yoshinaga T, Watanabe T, Sasaki S, Kamiyama S, Koizumi A (2004) The influence of time, sex and geographic factors on levels of perfluorooctane sulfonate and perfluorooctanoate in human serum over the last 25 years. J Occup Health 46:141–147
35. Jin Y, Saito N, Harada KH, Inoue K, Koizumi A (2007) Historical trends in human serum levels of perfluorooctanoate and perfluorooctane sulphate in Shenyang China. Tohoku J Exp Med 212:63–70
36. Kärrmann A, Ericson I, van Bavel B, Darnerud PO, Aune M, Glynn A, Lignell S, Lindström G (2007) Exposure of perfluorinated chemicals through lactation: levels of matched human milk and serum and a temporal trend, 1996–2004, in Sweden. Environ Health Perspect 115:226–230

37. Sundström M, Ehresman DJ, Bignert A, Butenhoff JL, Olsen GW, Chang SC, Bergman A (2011) A temporal trend study (1972–2008) of perfluorooctanesulfonate, perfluorohexanesulfonate, and perfluorooctanoate in pooled human milk samples from Stockholm Sweden. Environ Int 37:178–183
38. Human Biomonitoring Commission (2009) Umweltbundesamt, Referenzwerte für Perfluoroctansäure (PFOA) und Perfluoroctansulfonsäure (PFOS) im Blutplasma. Bundesgesundheitsbl Gesundheitsforsch Gesundheitsschutz 52:878–885
39. Wilhelm M, Angerer J, Fromme H, Hölzer J (2009) Contribution to the evaluation of reference values for PFOA and PFOS in plasma of children and adults from Germany. Int J Hyg Environ Health 212:56–60
40. Falandysz J, Taniyasu S, Gulkowska A, Yamashita N, Schulte-Oehlmann U (2006) Is fish a major source of fluorinated surfactants and repellents in humans living on the Baltic Coast? Environ Sci Technol 40:748–751
41. Haug LS, Thomsen C, Brantsaeter AL, Kvalem HE, Haugen M, Becher G, Alexander J, Meltzer HM, Knutsen HK (2010) Diet and particularly seafood are major sources of perfluorinated compounds in humans. Environ Int 36:772–778
42. Hölzer J, Göen T, Just P, Reupert R, Rauchfuss K, Kraft M, Müller J, Wilhelm M (2011) Perfluorinated Compounds in Fish and Blood of Anglers at Lake Möhne, Sauerland Area, Germany. Environ Sci Technol dx.doi.org/10.1021/es104391z.
43. Lundin JI, Alexander BH, Olsen GW, Church TR (2009) Ammonium perfluorooctanoate production and occupational mortality. Epidemiology 20:921–928
44. Sakr CJ, Leonard RC, Kreckmann KH, Slade MD, Cullen MR (2007) Longitudinal study of serum lipids and liver enzymes in workers with occupational exposure to ammonium perfluorooctanoate. J Occup Environ Med 49:872–879
45. Leonard RC, Kreckmann KH, Sakr CJ, Symons JM (2008) Retrospective cohort mortality study of workers in a polymer production plant including a reference population of regional workers. Ann Epidemiol 18:15–22
46. Fei C, McLaughlin JK, Tarone RE, Olsen J (2007) Perfluorinated chemicals and fetal growth: a study within the Danish National Birth Cohort. Environ Health Perspect 115:1677–1682
47. Apelberg BJ, Witter FR, Herbstman JB, Calafat AM, Halden RU, Needham LL, Goldman LR (2007) Cord serum concentrations of perfluorooctane sulfonate (PFOS) and perfluorooctanoate (PFOA) in relation to weight and size at birth. Environ Health Perspect 115:1670–1676
48. Hamm MP, Cherry NM, Chan E, Martin JW, Burstyn I (2010) Maternal exposure to perfluorinated acids and fetal growth. J Expo Sci Environ Epidemiol 20:589–597
49. Monroy R, Morrison K, Teo K, Atkinson S, Kubwabo C, Stewart B, Foster WG (2008) Serum levels of perfluoroalkyl compounds in human maternal and umbilical cord blood samples. Environ Res 108:56–62
50. Washino N, Saijo Y, Sasaki S, Kato S, Ban S, Konishi K, Ito R, Nakata A, Iwasaki Y, Saito K, Nakazawa H, Kishi R (2009) Correlations between prenatal exposure to perfluorinated chemicals and reduced fetal growth. Environ Health Perspect 117:660–667
51. Nolan LA, Nolan JM, Shofer FS, Rodway NV, Emmet EA (2009) The relationship between birth weight, gestational age and perfluorooctanoic acid (PFOA)-contaminated public drinking water. Reprod Toxicol 27:231–238
52. Fei C, McLaughlin JK, Lipworth L, Olsen J (2009) Maternal levels of perfluorinated chemicals and subfecundity. Hum Reprod 24:1200–1205
53. Human Biomonitoring Commission (2007) Umweltbundesamt, Derivation of human biomonitoring (HBM) values based on tolerable intake doses. Bundesgesundheitsbl Gesundheitsforsch Gesundheitsschutz 50:251–254, in German
54. Hays SM, Becker RA, Leung HW, Aylward LL, Pyatt DW (2007) Biomonitoring equivalents: a screening approach for interpreting biomonitoring results from a public health risk perspective. Regul Toxicol Pharmacol 47:96–109

Index

T.P. Knepper and F.T. Lange (eds.), *Polyfluorinated Chemicals and Transformation Products*, Hdb Env Chem (2012) 17: 169–172,
DOI 10.1007/978-3-642-21872-9,